Drawing the BORDERLINE

Artist-Explorers of the U.S.-Mexico Boundary Survey

THE ALBUQUERQUE MUSEUM

Drawing the Borderline:
Artist-Explorers and the
U.S.-Mexico Boundary Survey

AN EXHIBITION ORGANIZED BY
THE ALBUQUERQUE MUSEUM

Gray Sweeney, *Guest Curator,*
Arizona State University

John Grassham, *Curator of History,*
The Albuquerque Museum

EXHIBITION
CATALOGUE

PREFACE
Robert V. Hine

CONTRIBUTING AUTHORS

John Mack Faragher, *Arthur Unobskey*
Professor of American History,
Yale University

Oscar J. Martinez, *Professor,*
History Department, University of Arizona

Gray Sweeney, *Professor, School of Art,*
Arizona State University

Lucretia Hoover Giese, *Associate Professor,*
Art History, Rhode Island School of Design

Sam Truett, *Doctoral Candidate,*
History Department, Yale University

EDITOR
Dawn Hall

DESIGN AND TYPOGRAPHY
Christian Kenesson

The Albuquerque Museum
2000 Mountain Rd. NW
Albuquerque, NM 87104

LIBRARY OF CONGRESS CATALOG
CARD NUMBER: 96-83034

ISBN-0-8263-1752-9

COVER
Henry Cheever Pratt
View from Maricopa Mountain near the Rio Gila (detail), ca. 1855
33" x 48" • Oil on canvas
Loan courtesy of The Rosewood Corporation, Dallas, Texas

CONTENTS

Acknowledgements v

Foreword vii

Preface- A Border Redrawn • *By Robert V. Hine* viii

North, South, & West: Sectional Controversies and the U.S.-Mexico Boundary Survey • *By John Mack Faragher* 1
  Notes for Faragher essay 11

Surveying & Marking the U.S.-Mexico Boundary: The Mexican Perspective • *By Oscar J. Martinez* 13
  Notes for Martinez essay 22

Drawing Borders: Art and the Cultural Politics of the U.S.-Mexico Boundary Survey 1850-1853 • *By Gray Sweeney* 23
  Notes for Sweeney essay 75

Artist Collaborators A Surrogate Hand: Seth Eastman's for Bartlett's • *By Lucretia Hoover Giese* 79
  Notes for Giese essay 91

Borderlands and Border Places: The Historical Landscape of the U.S.-Mexico Border • *By Sam Truett* 95
  Notes for Truett essay 106

Color Plates 109

Exhibition List 145

THE ALBUQUERQUE MUSEUM

Acknowledgements

The Albuquerque Museum has had the most sincere cooperation from many individuals, collectors, museums and galleries for this catalog and exhibition and without their support this exhibition could never take place. At the John Carter Brown Library, Brown University, Providence, Rhode Island, Norman Fiering, Director and Susan Danforth, Curator of Prints greatly facilitated and expedited the loan of their Bartlett and Pratt drawings. At the Rhode Island School of Design, Providence, Rhode Island, Roger Mandle, Executive Director, Doreen Bolger, Director, and Maureen O'Brien, Curator of Prints, Drawings and Photographs, supported and assisted in the loan of their Seth Eastman paintings, as did board member, Al Morris. At Beinecke Library, Yale University, George Miles, Director, assisted in the loan of maps and historic sheet music. Dr. James K. Ballinger, Director of the Phoenix Art Museum and Michael Kominenski, Curator of American Art and Heather Northway, Registrar, assisted in the loan and facilitated the photography for the catalog. At the Amon Carter Museum, Rick Stewart, Director, and Melissa Thompson, Registrar, assisted in the loan of their Pratt painting. At the Texas Memorial Museum, University of Texas, Austin, Texas, Dr. William Reeder, Director, and Sally Baulch, Collections Manager assisted with the loan of their Pratt painting. At the Magoffin House State Historic Site, El Paso, Texas, Mary Kay Shannon, Assistant Park Superintendent, approved the loan of their Pratt portrait of James Magoffin. At the University Art Museum, University of New Mexico, Dr. Peter Walch, Director.

William Vareika Fine Arts in Newport, Rhode Island, quickly responded to our inquiry about a Pratt painting as did Abbot W. Vose, president of Vose Gallery, Boston; Jeffrey Cooley of Cooley

Gallery, Old Lyme, Connecticut and Mark Zaplin and Richard Lampert of Zaplin-Lampert Gallery, Santa Fe, NM. At the Kennedy Gallery, Inc. in New York City, Martha Fleischman, Director and Lillian Brenwasser, Vice President greatly assisted in our search for various Pratt paintings.

We extend our sincere appreciation to the private lenders who graciously loaned their paintings to the exhibition. They include Philip and Amy Alfeld; Arthur Phelan, Jr.; Philip Y. De Normandie; Dr. and Mrs. David Chase; Don Heald, New York City; William Blundin, New York City and the Rosewood Corporation, Dallas, Texas.

The Guest Curator of the exhibition, Gray Sweeney would like to acknowledge the following individuals at Arizona State University for assistance with preparing the exhibition and researching the catalog: Robert Barnhill, Vice President of Research and Creative Activities, Robert Barnhill; Ronald Barr, Associate Vice President; Richard Fill, Assistant to the Vice President; and J. Robert Wills, Dean of the College of Fine Arts; and Julie F. Codell, Director , School of Art.

Professor Sweeney would like to acknowledge the assistance of Merl M. Moore, Jr. who generously shared archival materials about Henry Cheever Pratt, and without whose assistance the project would have been difficult to complete. He would also like to acknowledge the assistance of Elisabeth Waldo Dentzel whose paintings by Pratt at the Phoenix Art Museum first suggested the potential importance of an investigation of the US-Mexico Boundary Survey. Arthur J. Phelan, Jr. assisted with archival research and provided valuable information about Pratt from his files. At the University of Texas, El Paso, Professors David Hackett and W. J. Timmons provided helpful information about Henry Cheever Pratt's painting of Coon's Rancho. Professor Joni Kinsey at the University of Iowa, and Professor Robert Mugerauer offered helpful comments on the manuscript. Professor Corine Schleif at Arizona State University offered suggestions and encouragement during the early stages of the investigation. In the art history program at Arizona State University Lou Morrision, Clarissa J. Howe and Allison Bertrand assisted by reading the manuscript. Professor Sweeney would especially like to acknowledge the support of his family, Karrie and James, during the preparation of the project.

FOREWORD

Drawing the Borderline: Artist-Explorers and the US-Mexico Boundary Survey is the first exhibition ever compiled in New Mexico on the visual images completed by the US-Mexico Boundary Survey commissioner, John Russell Bartlett, and artists Henry Cheever Pratt and Seth Eastman. Most of these drawings and watercolors in this exhibition date from 1850-1853 and were rendered on site in the course of the official survey. Thus, these works of art are considered primarily as cultural artifacts—documents of the expedition—and only secondarily as independent aesthetic creations. Although these images adhere to aesthetic conventions of the period and convey meanings inherent in those structures, they were not, except for some of the Pratt oils, made on speculation for the art market of nineteenth century America.

The Albuquerque Museum would like to thank Professor Gray Sweeney of Arizona State University for initiating this project and for his work over the last three years in helping to bring it to fulfillment. Professor Sweeney's contribution to locating oil paintings by Pratt was time-consuming, as most of the paintings not been gathered together since the 1870s. He also worked closely with the authors for this catalog.

Dr. Robert V. Hine first made historians and art historians aware of Bartlett and Pratt in his 1968 book, *Bartlett's West: Drawing the Mexican Boundary.* This ground breaking catalog is still considered a basic source on the subject. In reexamining the issues of the Boundary Survey, we sincerely appreciate the work in our catalog of Dr. Hine, Dr. John Mack Faragher, Arthur Unobskey Professor of American History, Yale University; Dr. Oscar Martinez, History Department, University of Arizona; Dr. Lucretia Hoover Giese, Professor of Art History, Rhode Island School of Design; and Yale doctoral candidate, Sam Truett.

James Moore, *Director*

John Grassham, *Curator of History*

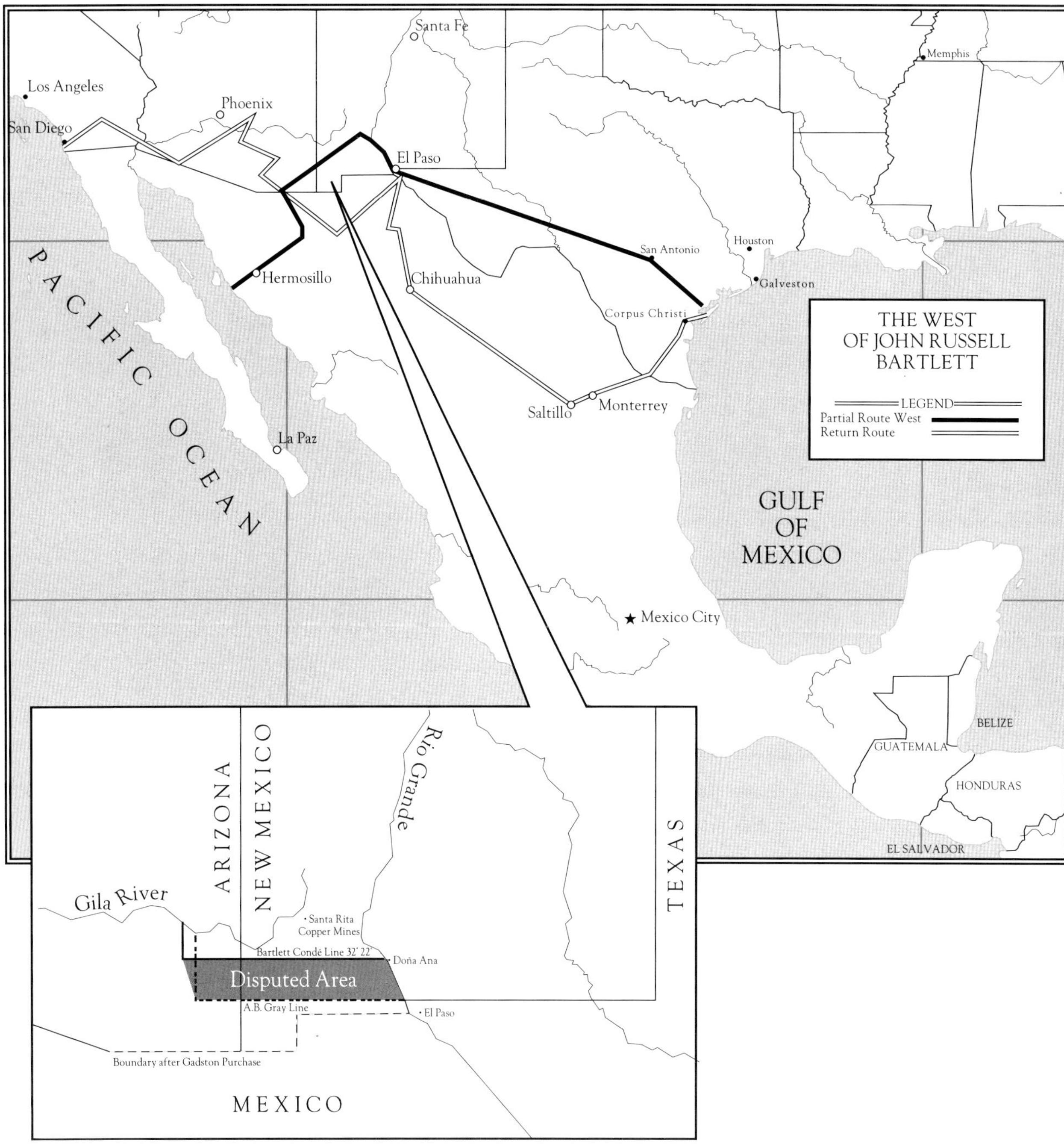
THE WEST
OF JOHN RUSSELL
BARTLETT
LEGEND
Partial Route West
Return Route
Santa Fe
Memphis
Los Angeles
Phoenix
San Diego
El Paso
San Antonio
Houston
Hermosillo
Chihuahua
Galveston
Corpus Christi
PACIFIC OCEAN
Saltillo
Monterrey
La Paz
GULF
OF
MEXICO
Mexico City
BELIZE
GUATEMALA
HONDURAS
EL SALVADOR
ARIZONA
NEW MEXICO
Rio Grande
TEXAS
Gila River
Santa Rita
Copper Mines
Bartlett Condé Line 32° 22'
Doña Ana
Disputed Area
A.B. Gray Line
El Paso
Boundary after Gadston Purchase
MEXICO

ROBERT V. HINE

PREFACE
A Border Redrawn

The border between the United States and Mexico poses issues far larger than its physical dimensions. A two thousand mile-long serpent stretched out across an arid countryside, it separates and attracts, impels and repels, and calls on supermen to put down the racism and demagoguery that rise around it. The birth of that border intrigues us; we are amused and appalled by beginnings that were intimately associated with a man named John Russell Bartlett, a literary scholar, bibliophile, and artist—not a classic heroic figure or superman—who was swept for a time from his native New England into the wild and turbulent frontier.

The history of the borderlands stretches back into the 16th century, long before "the border" hatched from that mythical serpent's egg of war. The border was identified as an area of study early in the 20th century by Herbert Eugene Bolton, the historian who first used the term "borderlands," and then was further refined by the contemporary southwestern historian David J. Weber in his writings. The ghostly landscapes of the borderlands evoke Indian civilizations that were already ancient when Cabeza de Vaca, Coronado, and Juan de Oñate walked over its sands. When John Russell Bartlett fixed his "magisterial gaze" on these same expanses, he was but one more "herald of empire," one human straw in the forces that have swirled across this desolate land for nearly five centuries.

Bartlett, Henry Cheever Pratt, and Seth Eastman are the three men who visually depicted the U.S.-Mexico Boundary Survey with their sketches, wood-cuts, drawings, watercolors, and paintings. They steadfastly shouldered many burdens: the physical hardships of accidents, injuries, and illness; the psychological conditioning of attractions and repulsions found in human relationships; and the pervasive prejudices expressed toward distinct groups of Mexican and Native American people. As official representitives of the Federal Government their art reflected the ideologies of pre-civil war America regarding the ultimate use of this desert landscape. Beyond the disparate American feelings of the time, xenophobia mingling with Manifest Destiny, glorious moments came to these artists during and after the boundary survey when they glimpsed unusual potential and saw previously closed avenues open. Their art became visionary as the artists saw beyond their own spectacles into future possibilities, and revised predominant images—the dreams as well as the reflections. Bartlett, Pratt, and Eastman accepted the challenge of expressing the excitement of a different and unusual terrain and produced a lasting volume of haunting images.

In 1962, I was asked by Thomas R. Adams of the John Carter Brown Library, Mitchell A. Wilder of the Amon Carter Museum, and Archibald Hanna of the Western Americana Collection of Yale University to write the John Russell Bartlett story. They felt the art produced during Bartlett's boundary survey of 1850 to 1853 needed to be brought to the attention of the American public, especially to those with a historical consciousness. Like most of my history colleagues, I was only fleetingly familiar with the name Bartlett. At that time, when the boundary survey was mentioned the name William Emory came most readily to mind. Leafing through Bartlett's, Pratt's, and Eastman's striking works of art that were preserved in the John Carter Brown Library, and reading accounts of the survey, I became more and more excited about the prospect of a Bartlett project. The historical messages conveyed, and the warmth and fascination of the human beings involved held the promise of inspiration. The only real study of the work of Commissioner Bartlett at that time, aside from entries in biographical dictionaries, was Odie Faulk's introduction to a new edition of Bartlett's *Personal Narrative of Explorations and Incidents in Texas, New Mexico, California, Sonora, and Chihuahua.* I agreed to write the book *Bartlett's West: Drawing the Mexican Boundary*, which was published in 1968. Curiously, not much has been written since the 60s. Now, more than thirty years after my initial viewing of the boundary survey art, The Albuquerque

Museum has mounted the exhibition "Drawing the Borderline," guest-curated by Gray Sweeney, thus bringing the survey into focus again.

This exhibition and accompanying catalog do far more than retell the tale. In these intervening thirty years, America, its historians, and its art historians have followed divergent pathways—comparisons and analogies, metaphors, signifiers, and tropes. The political history of the American West emphasizes conquests, exploitation, and class-consciousness. Our horizons now have expanded in all directions, encompassing race, gender, and the environment.

Modern enthusiasts of the art of Bartlett, Pratt, and Eastman will again be exposed to the boundary survey years. Gray Sweeney and Lucretia Hoover Giese present the viewpoint of the art historian. Oscar Martinez explores the story as it appeared from the Mexican side of the border. John M. Faragher recounts in fresh terms the historical forces that impelled the survey. And Sam Truett adds the dimensions of the border environment and ecology.

Thirty years is a long time to wait to absorb another look at the Southwest as it was in the 1850s. The resultant exhibition and catalog, however, merit the wait.

JOHN MACK FARAGHER

NORTH, SOUTH, AND WEST

Sectional Controversies and the U.S.-Mexico Boundary Survey

The boundary between the United States and Mexico has long been troublesome. The proximate cause of the Mexican War was a border dispute, and when peace was restored in 1848 one of the first tasks was redrawing the boundary, an assignment given to the U.S.-Mexico Boundary Survey. When the international commission, led by Commissioners John Russell Bartlett of the United States and Pedro García Condé of Mexico, adjourned after five years of work in 1853, however, the two countries stood once again on the brink of armed conflict over location of the borderline.

At the time, most of the blame fell on United States Boundary Commissioner John Russell Bartlett. Critics lambasted Bartlett's management, or mismanagement of the work. He authorized the spending of approximately $500,000 during the survey, only about $100,000 of which actually went for running and marking the boundary; the rest was wasted, or spent on the Commissioner's protracted travels with his entourage through countrysides far removed from the boundary itself. U.S. government and territorial leaders condemned him for the compromise agreement he made with his Mexican counterpart. Even President Millard Fillmore, the leader of Bartlett's own Whig party, disavowed his work. According to historian Odie B. Faulk, Bartlett was a failure as Boundary Commissioner.[1]

The failure of the boundary survey itself had deeper causes. There were Americans prepared to object to nearly any boundary drawn with Mexico. For example, young Stephen A. Douglas, Democratic congressman from Illinois, declared on the floor of the House of Representatives in 1845 that he "would blot out the lines on the map which now marked our national boundaries on this continent, and make the area of liberty as broad as the continent itself." An empire for liberty: it was an old American ideal. But never before had Americans asserted it so boldly, or with such disregard for the fences of their neighbors. "Acquisitions of territory in America, even if accomplished by force of arms, are not to be viewed in the same light as the invasions and conquests of the States of the old world," asserted the editor of the *New York Morning News*. "Our way lies, not over trampled nations, but through desert wastes."[2]

Manifest Destiny

The *Morning News* was one of the dozens of cheap American newspapers — the "penny press" — that revolutionized the dissemination of information in the 30s and 40s. Its editor, John L. O'Sullivan, a chief propagandist for the Democratic Party, coined one of the most famous phrases in American history when he insisted on "our manifest destiny to overspread the continent allotted by Providence for the free development of our yearly multiplying millions." Over the years many Americans had called attention to America's "multiplying millions," a phenomenon English economist Thomas Malthus described as "a rapidity of increase probably without parallel in history." Congressman Andrew Kennedy of Indiana put it in prosaic terms:

> *Go to the West and see a young man with his mate of eighteen; and [after] a lapse of thirty years, visit him again, and instead of two, you will find twenty-two. That is what I call the American multiplication table.... How long, under this process of multiplication, will it take to cover the continent with our posterity, from the Isthmus of Darien to Behring's straits?*

Americans would "overspread the continent" simply by doing what came naturally. But O'Sullivan and other imperial ideologues took this thought a step further, arguing that American fecundity amounted to nothing less than a revelation of "the manifest design of Providence in regard to the occupation of this continent."[3]

Pompous rhetoric of this kind did not bubble up from the folk. It was a self-conscious creation of propagandists with something specific in mind. They sought to uncouple the politics of expansion from the sectional controversy over slavery. A majority of 19th century Americans probably did believe in their right to expand unimpeded across the continent, but there was serious sectional division over whether they should be allowed to take their slaves along. The public controversy over the expansion of slavery first erupted in 1820 when Missouri applied for admission to the union as a slave state. Congress finally was able to sweat out a compromise, agreeing that slavery would be "forever prohibited" in any subsequent states carved from the Louisiana Purchase territory. This political settlement laid the foundation for the Democratic sectional coalition that swept Andrew Jackson to the presidency in 1828. At the end of Jackson's second term the controversy resurfaced when Texas applied to join the union. One of the issues that had driven the Texans to rebellion was their commitment to slavery despite the anti-slavery clause in Mexico's republican constitution of 1824. When Jackson's political opponents joined with abolitionists to oppose the annexation of a slave state the Democrats decided to shelve the issue.

After the Whig party captured the White House with William Henry Harrison's bombastic "log cabin and hard cider" campaign of 1840, however, Democratic strategists determined to play the expansionist card for all it was worth. What they needed was a political discourse that emphasized the benefits of expansion while taking the focus off slavery, and they believed they had found it in "manifest destiny." Focusing on natural, immutable, and "providential" forces, manifest destiny discourse managed to avoid the discomfiting fact of slavery, and to ignore both the purposeful federal policy and the ruthless use of military power that would be necessary to conquer the continent. It was the upside-down world of ideology — where what is *manifest* is mystification, what is *latent* is substance. Manifest destiny is a classic American case study in the uses of ideology — propaganda formulated for public consumption that rationalizes, excuses, or otherwise masks the true state of affairs in the interests of political power.

Historians seem more convinced of the power and scope of manifest destiny ideology than were voters at the time. If the presidential contest of 1844 was a referendum on expansion, as the Democrats claimed it was, its outcome suggested a nation of closely divided opinions. James K. Polk of Tennessee failed to win an absolute majority (with just 49.6% of the national vote), although he carried the election with fifteen of twenty-six states. Had his opponent,

Henry Clay of Kentucky, gathered only five thousand additional ballots in the state of New York — where a third party anti-slavery ticket drew the support of more than fifteen thousand voters — he would have won in the electoral college. It was remarkably close, yet the Democrats insisted they had won a mandate for a program of national expansion. Presidential elections, of course, are not about determining public opinion, but about capturing political power and putting it to use. A mandate is what one makes of it, and the Democrats were about to make a western empire with theirs.

American Blood on American Soil

For decades American filibusters had sought to subvert the northern provinces of Mexico while American presidents simultaneously sought to purchase them. As far as the Mexicans were concerned, it amounted to gringo imperialism either way. After the Texas rebellion, relations between the two countries grew steadily worse. Mexico's president, Antonio López de Santa Anna, formally notified the American government that annexation would be "equivalent to a declaration of war," and diplomatic observers in Washington were virtually certain that both annexation and war would come soon.[4]

Figure 1

Matamoros Grande March, 1846

Reproduced with permission from the Yale Collection of Western Americana, Beinecke Rare Book & Manuscript Library.

The "Matamoros Grande March," was issued by a Brooklyn publisher of sheet music when the United States declared war against Mexico in 1846. American soldiers marching off to battle were reported to have sung the following lines to the tune of "Ol' Dan Tucker":

"We're on our way to Matamoros
On our way to Matamoros
On our way to Matamoros
And we'll conquer all before us!"

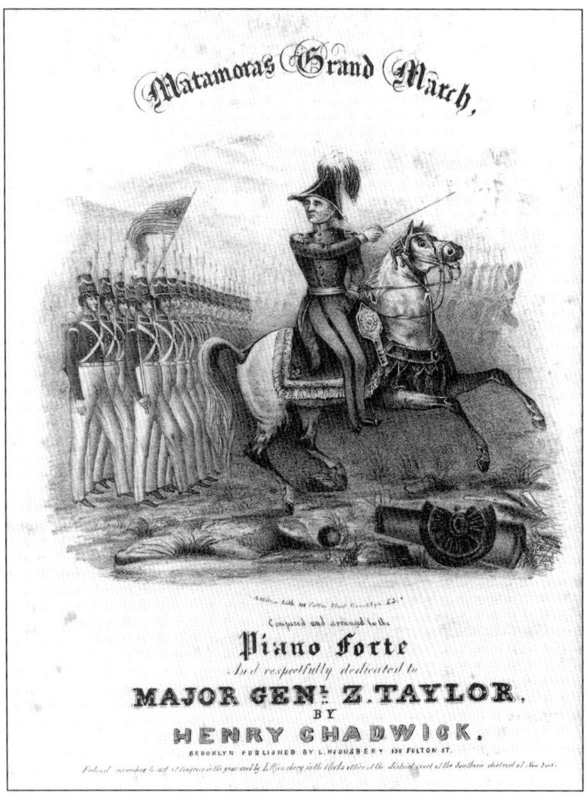

Just before Polk assumed office in March 1845, Congress voted to admit Texas to the Union. Mexico immediately severed diplomatic relations. The two countries negotiated, but to no avail. The Mexicans were angry, although there was no sign they were about to begin a war to reclaim a province they had lost ten years before. Polk, however, was spoiling for a fight. He had already announced to his Cabinet that the acquisition of California — with its fine Pacific ports — was the prime goal of his presidency. To force the issue, in early 1846 he ordered General Zachary Taylor to march an American army beyond the Nueces River, the traditional southern boundary of Texas, and toward the Rio Grande. Texas and Mexico had disputing claims to this strip of territory. The Texans pointed to the boundary agreed to by President Santa Anna after his capture at the conclusion of the rebellion in 1836. The Mexicans protested that Santa Anna had signed with a gun pressed against his temple, that the area had never been considered a part of Texas, and was populated exclusively by Mexicans. The Mexicans warned Taylor to retreat, but he refused, and in April there was a skirmish in which both sides took casualties. Polk now had his rationale for war. "Mexico has passed the boundary of the United States," he announced in his war message to Congress, "has invaded our territory, and shed American blood upon the American soil." Congress declared war on 13 May 1846. Believing that victory would not take long, the House of Representatives issued a call for fifty thousand short-term volunteers and appropriated $10 million to get the job done.[5]

By the end of 1846 American forces were in control of most of northern Mexico — including New Mexico and California — but the Mexicans made no signs of surrender. In a move designed to end the war, the United States landed troops at the Gulf Coast port of Vera Cruz and struck inland for Mexico City in May 1847. The fighting was intense and prolonged. Mexican losses at the front were followed by guerrilla actions in the rear, harassing supply lines and slowing progress. The Americans retaliated with outrages against civilians. According to General Winfield Scott, commander of the American forces, his troops "committed atrocities to make Heaven weep and every American of Christian morals blush for his country," including "murder, robbery and rape of mothers and daughters in the presence of tied-up males of the families." It was September before the Americans stormed Chapultepec, the palace of the viceroys on the outskirts of Mexico City, then seized the capital itself. The war was finally over, but it had cost the United States nearly $100 million. By official and conservative government reckoning, it also took the lives of some thirteen thousand Americans and at least twenty thousand Mexicans.[6]

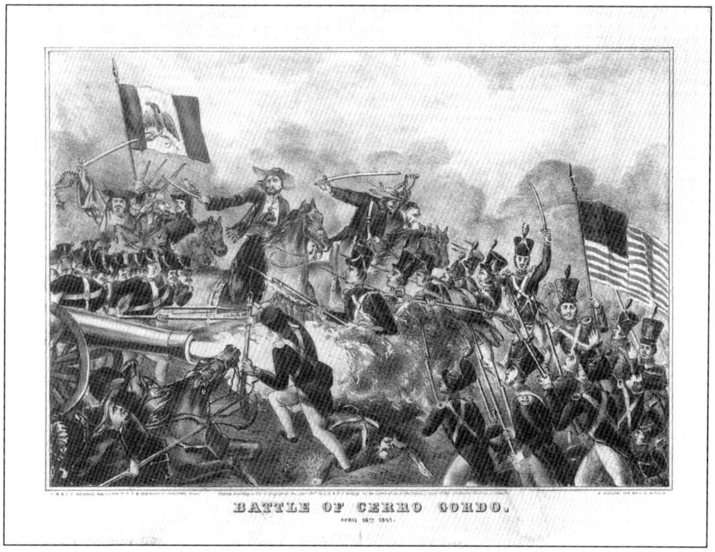

Figure 2

Battle of Cerro Gordo

Reproduced with permission from the Yale Collection of Western Americana, Beinecke Rare Book & Manuscript Library.

In this popular print, the American army sweeps up Telegraph Hill into the Mexican lines at the Battle of Cerro Gordo, 18 April, 1847, unusual for the detail with which it depicts the fenders, including irregular cavalry. Later guerrilla fighting led to atrocities on both sides. The Mexican War cost the lives of 13,000 Americans and at least 20,000 Mexicans.

The Politics of Conquest

The Democrats had anticipated that a patriotic war for territorial expansion would unify the country, but the rhetoric of "manifest destiny" could not paper over political and sectional divisions. Some Americans considered the entire Mexican War immoral and unworthy of the nation. Anti-slavery activists were in the forefront of this attack. They charged that the South had fomented the war to extend the institution of slavery. The war made him "sick at heart," declared the abolitionist Frederick Douglass, himself an escaped slave. "The spirit of slavery reigns triumphant throughout all the land. Every step in the onward march of political events is marked with blood — innocent blood; shed, too, in the cause of slavery." But opposition to the war was broader than the anti-slavery constituency alone. In the Senate the Whig minority proposed a "no territory by conquest" resolution that lost by the narrow margin of 29 to 24.[7]

Soon the Democrats too were dividing into hostile camps. Only a few months into the fighting, Congressman David Wilmot of Pennsylvania introduced an amendment to an appropriations bill providing that "neither slavery nor involuntary servitude shall ever exist" in territory acquired from Mexico. Wilmot was no anti-war Whig, but a Democrat who believed his party had been captured by Southern interests. Neither was he an abolitionist. Believing that "the negro race already occupy enough of this fair continent," he wanted to preserve the West for "the sons of toil, of my own race and own color." The House passed the Wilmot Proviso, but Southerners in the Senate blocked it, declaring that they would not be shut out of the enjoyment of the spoils.[8]

Meanwhile, President Polk was flirting with the complete annexation of Mexico. He sent peace envoy Nicholas Trist to Mexico in early 1847 with instructions to secure the Rio Grande boundary for Texas, and to arrange for the cession of New Mexico and California. With the fall of Mexico City, however, forces within the administration began to press for the annexation of all Mexico. The Mexicans were economically backward, they argued, their political institutions were chaotic, and consequently they deserved to lose jurisdiction over regions that Americans could develop more efficiently. Secretary of State James Buchanan reported to the cabinet that Mexicans dreaded the day when American troops would withdraw from Mexico City, and Treasury Secretary Robert J. Walker commissioned a study of the fiscal implications of absorbing Mexico.

The "all-Mexico" movement raised further sectional divisions. Southern spokesmen trembled at the implications of incorporating "the colored and mixed-breed" population of Mexico. "We make a great mistake, sir," John C. Calhoun of South Carolina argued in the Senate, "when we suppose that all people are capable of self-government." More than half of Mexico's residents were Indians, and the rest "impure races, not as good as the Cherokees or Choctaws." The Spanish had made the "fatal error of placing these colored races on an equality with the white race," Calhoun lectured his colleagues, and thus "destroyed the social arrangement which formed the basis of society." And "yet it is professed ... to erect these Mexicans into a Territorial Government, and place them on an equality with people of the United States. I protest utterly against such a project."[9]

Amid this political turmoil Trist returned to Washington in early 1848 after negotiating a treaty based on Polk's earlier aims. The accomplished fact of peace stunned both the "no territory" and the "all-Mexico" forces, and a centrist majority in Congress ratified the agreement in March. The Treaty of Guadalupe Hidalgo confirmed the annexation of Texas to the United States and set the Rio Grande as its southern boundary. Mexico ceded to the United States upper California (but not Baja) and the province of New Mexico (covering the present-day states of New Mexico, Arizona, Utah, and Nevada). The United States paid Mexico $15 million and assumed responsibility for nearly $4 million of private American claims against the Mexican government. The two countries agreed to establish a joint commission to redraw the boundary between them.

Too Far North

The unsettled political controversies of the war would dominate the work of the boundary survey and its Commissioners. As first Commissioner the Democrats named John B. Weller, a colonel of militia in the Mexican War and a former Ohio congressman; he and his Mexican counterpart began the first stage of the survey. But running as a Whig and a war hero, General Zachary Taylor triumphed in the presidential contest of 1848, and his new administration removed Weller on trumped up charges and named as his replacement John C. Frémont, late of the Army Corps of Topographical Engineers, famous for his western explorations and his leadership of the "Bear Flag Rebellion" of American settlers in California during the war. After several months of indecision, however, Frémont chose instead to accept the draft of the California legislature to become one of the state's first two senators. It was then that the Whigs turned to John Russell Bartlett, who had been seeking a patronage appointment within the new administration.

Bartlett had neither diplomatic nor topographical experience, but he was a loyal Whig with influential political connections. For nearly fifteen years he had run a Manhattan bookstore, a popular haunt of intellectuals and academics, and had become something of an expert in the new discipline of ethnology. In 1847 the American

Ethnological Society published his *The Progress of Ethnology,* which Bartlett offered as an account of his colleague's efforts "towards extending our knowledge of the globe, particularly with reference to its geography, and to those nations whose history is imperfectly known." He was also fascinated with language, and was fluent in Spanish and French. His *Dictionary of Americanisms,* first published in 1849, was a pioneer study of colloquial American speech that highlighted Bartlett's keen awareness of the regional and ethnic diversity of the country. To the charge, for example, that American letters could never hope to reach English standards, he predicted that "the heterogeneous elements of which our society is composed will result in the production of a style and a literature which will also have their beauties and merits, although fashioned after a somewhat different model." He had impressive intellectual credentials.[10]

Bartlett and his party of more than a hundred co-workers and employees arrived in Paso del Norte, Chihuahua, in November 1850, where he soon met his Mexican counterpart, General Pedro García Condé, a native of Sonora who had risen through the ranks of the Mexican army to become Secretary of War and Marine during the recent conflict with the United States. Both men were in their mid-forties and had intellectual interests; they got along famously. As historian Robert Utley notes in his study of the boundary survey, Bartlett and Condé "set an example of understanding, teamwork, and good humor that might well have been emulated in Washington and Mexico City."[11]

Almost immediately, however, the two men encountered a significant impediment to their work. The Treaty of Guadalupe Hidalgo had provided a verbal description of the new boundary. It was to follow the course of the Rio Grande to its intersection with the southern boundary of New Mexico; continue west along that line to its termination, then north to the nearest branch of the river Gila; down the Gila to its mouth on the Rio Colorado; and finally west to the Pacific Coast. The official version of the treaty also included a copy of J. Disturnell's 1847 commercial map of the United States and Mexico, which showed the southern boundary of New Mexico running a few miles north of Paso del Norte. Once on the ground the Commission's astronomers found that Disturnell's map located Paso del Norte 34 miles north and 130 miles east of its actual geographic coordinates. The question was, should the Commissioners run New Mexico's southern boundary slightly north of the actual location of Paso del Norte, or follow the incorrect coordinates of the map? Bartlett predictably argued for the first possibility, which would result in more territory gained for the United States, Condé the second. Bartlett's official instructions had urged him to maintain a conciliatory relationship with the Mexicans, and that was his inclination anyway. He proposed a compromise. He would agree to the more northern latitude for the boundary if Condé in turn would allow the line to run west for three degrees of longitude. Essentially this split the difference between the positions, and after consulting his team, Condé agreed. The treaty had empowered the Commissioners to make binding agreements on just such particulars. As long as the commissioners and the head surveyors of both nations signed off, the compromise would not require the ratification of their respective governments.

The official American surveyor, however, refused to add his signature to the agreement. Andrew Belcher Gray was a Virginian who had worked previously for the Republic of Texas. He protested that the compromise line was "too far north," and that the Americans should hold out for the most possible territory. At stake was an arid strip of some six thousand square miles that Bartlett claimed could "never be inhabited." But that was not the issue, Gray argued. Now that the United States was a continental nation, it was only a matter of time before a great transcontinental railroad linked the two coasts. Indeed, Bartlett's instructions included assessing the possibility of a railroad line through the border region. Gray argued that this precise strip of desert was "the great gateway to the Pacific." A railroad connecting Texas and the South with California would, in his opinion, require a right of way "over the disputed district to the Gila." What was at stake in the disagreement between Bartlett the Yankee and Gray the Virginian, writes Robert V. Hine, "was the whole national debate between North and South." The sectional dispute over the appropriate route for the transcontinental railroad would preoccupy Congress over the next two years, and result in a series of extensive surveys to assess several competing routes across the West.[12]

Mexico Will Poison Us

The work of surveying the other sections of the boundary continued, and Bartlett embarked on the travels that would provide the subjects for the drawings and paintings of the expedition, as well as the substance of his *Personal Narrative of Explorations and Incidents in Texas, New Mexico, California, Sonora, and Chihuahua.* But

Surveyor Gray forwarded his protest to Washington and soon Congress was investigating Bartlett's compromise. The Commissioner was pilloried by the Democrats as a man who cowered before the Mexicans, and in 1852 Congress passed a resolution demanding that not a penny more be spent on the boundary survey until the line was shifted south. In the presidential election of 1852 the Democrats captured the presidency by accusing the Whigs of giving away what had been won with American blood. But even Whig President Millard Fillmore (succeeding Taylor, who died in 1850) concluded that Bartlett had acted improperly and notified Mexico that the boundary would have to be renegotiated. With his honor on the line, Bartlett had no choice but to terminate his activities on the boundary survey and return home.

The rejection of Bartlett's work by his own government set the stage for armed confrontation on the New Mexico boundary. In Washington the new administration selected James Gadsden, a South Carolina railroad promoter, to go to Mexico to negotiate with President Santa Anna for a new boundary. Gadsden was instructed to demand a huge slice of Mexico's northern provinces, but be willing to settle for the controversial strip Bartlett had "given away," as well as additional territory south of the Gila River. "The valley must belong to the United States by an indemnity," Santa Anna remembered Gadsden warning him, "or we will take it." The Mexicans refused to consider the larger cession but agreed to the lesser. As Santa Anna put it, he "recognized that it was impolitic to refuse." The United States paid Mexico another $10 million for the additional territory and war was avoided.[13]

Americans were soon killing each other over the implications of expansion. When the Senate took up the Gadsden Treaty in 1854 it was in the midst of the debate over the status of slavery in the new territory of Kansas, and the close vote to ratify the treaty echoed the sectional politics that were already threatening to tear the nation apart. Within a year Kansas was bleeding, beginning a Civil War that would soon consume the country. Such were the political passions that insured the failure of Bartlett's boundary survey. "The United States will conquer Mexico," Ralph Waldo Emerson had predicted during the Mexican War. "But it will be as the man swallows the arsenic which brings him down in turn. Mexico will poison us."[14]

NOTES

North, South, and West: Sectional Controversies and the U.S.-Mexico Boundary Survey

1. Odie B. Faulk, *Too Far North ... Too Far South* (Los Angeles: Westernlore Press, 1967).

2. Frederick Merk, *Manifest Destiny and Mission in American History: A Reinterpretation* (original 1963; Cambridge: Harvard University Press, 1995), 25n,28.

3. Merk, *Manifest Destiny*, 27-28,29; John L. O'Sullivan, "Annexation," *Democratic Review* 17 July-August (1845):5-10; Thomas Malthus, *Essay on the Principle of Population*, ed. Michael P. Fogarty (New York: 1958), 1:305-06.

4. Richard B. Morris, ed., *Encyclopedia of American History: Bicentennial Edition* (New York: Harper & Row, 1976), 225.

5. William MacDonald, ed., *Select Documents Illustrative of the History of the United States, 1776-1861* (New York: Macmillan, 1905), 352.

6. Carey McWilliams, *North From Mexico: The Spanish-Speaking People of the United States* (original 1948; New York: Greenwood, 1968), 102-03.

7. Sean Wilentz, ed., *Major Problems in the Early Republic, 1787-1848* (Lexington, MA: D.C. Heath, 1992), 538.

8. James M. McPherson, *Battle Cry of Freedom: The Civil War Era* (New York, 1988), 55.

9. Merk, *Manifest Destiny*, 162.

10. John R. Bartlett, *The Progress of Ethnology, An Account of Recent Archaeological, Philological and Geographical Researches in Various Parts of the Globe, Tending to Elucidate the Physical History of Man* (New York: American Ethnological Society, 1847), 3; Bartlett, *A Dictionary of Americanisms* (New York: Bartlett & Welford, 1849), xxvii.

11. Robert M. Utley, *The International Boundary, United States and Mexico: A History of Frontier Dispute and Cooperation, 1848-1963* (Santa Fe: United States Department of the Interior, National Park Service, Southwest Region, 1964), 6.

12. Faulk, *Too Far North*, 75-76; Robert V. Hine, *Bartlett's West: Drawing the Mexican Boundary* (New Haven: Yale University Press, 1968), 40.

13. Faulk, *Too Far North*, 131,138.

14. Bernard DeVoto, *The Year of Decision, 1846* (Boston: Houghton Mifflin, 1942), 492.

OSCAR J. MARTINEZ

SURVEYING AND MARKING THE U.S.-MEXICO BOUNDARY

The Mexican Perspective

The surveying and marking of the boundary between the United States and Mexico, which began in 1849 and ended in 1856, constitutes the final chapter in a long conflictive relationship between the two nations over border delimitation. Mexico emerged from this struggle dispossessed of half of its national domain, while the United States expanded its holdings by almost a third. Texas fell to the *norteamericanos* through annexation in 1845, and New Mexico, Arizona, California, Nevada, Utah, and parts of Wyoming, Colorado, Kansas, and Oklahoma were acquired as part of the Treaty of Guadalupe Hidalgo in 1848; finally, the 1854 Gadsden Purchase added portions of Sonora and Chihuahua to the U.S. land mass.

Mexico's devastating loss of territory is remembered by its people as one of the most critical junctures in their history. At the time many Mexicans expressed disgust with their national leaders, heatedly debated the effectiveness of governmental structures, and somberly questioned the ability of their fragmented society to achieve national unity. Much of the wrath and bitterness spawned by the humiliation of the 1840s, however, was directed at the United States, viewed by Mexicans as an aggressive nation bent on conquest and domination of its neighbors. The arrogance and expression of racial superiority shown by Americans elicited deep resentment.[1]

In the late 1840s and early 1850s Mexican suspicion and distrust of the United States continued to grow as Americans attempted to secure transportation privileges in the strategic Isthmus of Tehuantepec, as filibusters based on U.S. soil repeatedly invaded Mexico's northern provinces, and as Apache and Comanche Indians continued to plunder frontier settlements in Sonora and Chihuahua. Mexican politicians and diplomats proceeded with extreme caution when conducting negotiations with American officials. Tact was required in order for Mexico to prevent more territory loss to its land-hungry neighbor and to avoid suffering additional violations of its national sovereignty.

Under such conditions, members of the Mexican Commission of the U.S.-Mexico Boundary Survey, assigned to identify the exact location of the political line dictated by the 1848 treaty, faced a particularly difficult challenge. The work had to be conducted jointly with American counterparts, some of whom proved difficult to deal with because of their ethnocentric and expansionist views. For the Mexican survey commissioners the task at hand demanded suppression of personal resentments and painful memories. It also required considerable patience, for there would be continuous delays in the field brought on by American political maneuverings and frequent changes in personnel on the U.S. Commission. Chronic shortages of funds and spartan living conditions on the desert would take their toll on both Mexicans and Americans, and, as the work progressed, both delegations would see their diplomatic skills tested. As the party bargaining from the weaker position, however, the Mexican contingent would have to overcome more hurdles and make greater adjustments than the Americans.

Isleta, Socorro, and San Elizario

The first dispute over the location of the boundary involved Isleta, Socorro, and San Elizario, three towns near Paso del Norte (present-day El Paso-Ciudad Juárez) with a combined population of nearly three thousand, set in the midst of very fertile agricultural lands. At the time of the signing of the Treaty of Guadalupe Hidalgo these communities were situated on an island surrounded by two channels of the Rio Grande. Before 1828 the river ran exclusively north of the towns, but sometime during the next six years a second channel formed to the south. The Treaty of Guadalupe Hidalgo specified that wherever the river split up into two or more channels, the deepest one would constitute the boundary, and thus the question arose whether the three settlements, and the surrounding valuable lands, should remain in Mexico or be engulfed by the United States.[2]

Without waiting for the boundary commissions from both nations to render a decision on the issue, as stipulated in the treaty, U.S. troops took possession of the towns and expelled the Mexican authorities in January 1849, declaring the southern channel the deeper one and therefore the new boundary.[3] When the mayor of Paso del Norte objected to the U.S. occupation to the nearest American magistrate (in Frontera, New Mexico), the magistrate arrogantly stated that no one had complained about the change in jurisdiction and asked the mayor to present property titles from the three towns. The magistrate further warned the Mexican mayor about the possibility of a disturbance. In expressing concern about these developments, the governor of Chihuahua informed the Mexican Secretary of Foreign Relations that it would be easy to drive out the invaders, but he counseled restraint because that course of action might lead to more serious confrontation.[4]

Vehement diplomatic protest by the Mexican government prompted the United States to withdraw its soldiers, but a few years later the commissioners confirmed the southern channel as the legal boundary, thus ending the controversy. Had the commissioners investigated the matter in 1848 or 1849, Mexico would have retained the towns because, according to Mexican historian Luis G. Zorrilla, at the time the northern channel was the deeper one.[5] In any case, willing or not, several thousand Mexicans living in Isleta, Socorro, and San Elizario were incorporated into the United States. Under the newly imposed U.S. legal system, these towns lost their *ejidos*, or communal landholdings, as did San Lorenzo and Cenecú, two settlements situated south of the river that had *ejidos* north of the river. Land claims submitted to American authorities by Mexican citizens yielded no results.[6] To Mexico, the invasion of these towns demonstrated the disregard the U.S. government had for procedural stipulations contained in the Treaty of Guadalupe Hidalgo. The Isleta-Socorro-San Elizario episode, says Mexican historian Angela Moyano Pahissa, replicated the expansionist process previously seen in Texas: "invasion of coveted land, change of officials in charge, pretense of holding a referendum, threats, claims of innocence on the part of federal authorities, and annexation of territory."[7]

Frustrated Compromise

The second and more important controversy pertained to the location of the boundary westward from the Rio Grande north of Paso del Norte. Following the uneasy but largely trouble-free determination of the initial location of the boundary on the Pacific and the point of junction at the Colorado and Gila rivers,[8] the commissioners agreed to meet at Paso del Norte in November 1850, to begin the task of drawing the line from the Rio Grande to the Gila River.

At Paso del Norte, their toughest challenge was to interpret the vague assertion in the treaty that the dividing line should be placed "north of the town called Paso;" the problem was complicated by an error in the Disturnell Map of 1847, a key document used in the treaty negotiations, which placed Paso del Norte 34 miles north and 130 miles east of its true location. From Mexican Boundary Commissioner Pedro García Condé's perspective, which reflected the official view in Mexico City, the line westward from the Rio Grande had to begin at 32° 30' north latitude since Disturnell had placed the southern boundary of New Mexico at that point. That interpretation clashed with American wishes to begin the line much closer to Paso del Norte, at about 31° 52' north latitude, emphasizing the need to correct the error made by Disturnell.

García Condé proposed a line from the Rio Grande to the Gila River that would leave the Santa Rita del Cobre mines and the Mesilla Valley in Mexico; U.S. Boundary Commissioner John Russell Bartlett countered with a line that would place the valuable real estate in the United States and would assure a much-desired pathway for a railroad route to the Pacific. The two sides compromised by agreeing on a boundary that would leave the Mesilla Valley in Mexico but place the Santa Rita mines in the United States and include enough territory for the U.S. railroad route, at 32° 22' north latitude. García Condé considered this agreement a great victory because the primary interests of Mexico, retaining fertile agricultural lands and the important towns of Mesilla and Paso del Norte, had been accomplished.[9] Bartlett also saw the compromise as a triumph because he knew the land obtained by the United States had rich deposits of gold and silver, and much of the territory retained by Mexico had little value.[10]

Subsequent events unraveled the work of the two commissioners. Lieutenant Amiel Weeks Whipple, the surveyor ad interim for the U.S. Commission, had just started to trace the new line when the army reassigned him. Eight months passed before permanent surveyor Andrew B. Gray assumed his duties following an extended illness. The resulting delays "infuriated" García Condé. "The increased expenses I am incurring in this desert," he wrote to Bartlett, "and the loss of time could lead to serious consequences. I am telling you how I feel and I am reporting it to my government." Embarrassed, Bartlett apologized.[11]

Much to the dismay of both García Condé and Bartlett, Surveyor Gray disapproved of the 32° 22' north latitude line, and refused to sign the compromise document; Lieutenant Colonel James D. Graham, then the principal astronomer, and Whipple himself, also repudiated the García Condé-Bartlett pact. Lacking the support of these key individuals, Bartlett suspended surveying activities, and asked García Condé to do the same. García Condé declined, and the Mexican Commission continued to trace the line toward the Gila River in accordance with the provisions of the compromise. But García Condé's efforts bore no fruit, for Gray's objections succeeded in bringing the drawing of the boundary in the Mesilla strip to a halt. "The biggest fears of the Mexican government had been realized," wrote Joseph Richard Werne. "The United States adopted and staunchly held on to Gray's interpretation of Article V of the Treaty of Guadalupe Hidalgo; the subsequent dispute gave that country sufficient diplomatic pretexts to demand a new treaty...."[12]

García Condé and his associates on the Mexican Commission must have felt considerable anger, frustration, and resentment at the unfavorable turn of events. Dealing with Gray must have been especially difficult, too, because of his background. In the 1830s, the U.S. surveyor assisted in the drawing of the eastern-northern boundary of the "Republic of Texas;" at the time, Mexicans viewed Texas as a renegade province that had fallen into the hands of Anglo American separatists, mercenaries, and pirates. Gray also rode with the Texas Rangers, a notorious law enforcement agency well-known for persecuting Mexicans and Indians.[13]

Throughout his time on the boundary survey, García Condé and his superiors in Mexico City had been annoyed with frequent delays prompted by changes in personnel on the U.S. Commission, and by the apparent lack of interest shown by the Americans to expeditiously identify the location of the new boundary, choosing instead to conduct explorations elsewhere, searching for the best terrain to build roads, canals, and railroads, and assessing the potential of the region's natural resources.[14] "It is not farfetched to assume that the suspension of activity or slowness in drawing the boundary had the approval of the government in Washington," wrote Zorrilla.[15] The implication in Zorrilla's comment was that the United States intentionally slowed down and complicated the process in an effort to buy time to learn more about the region and subsequently extract from Mexico as much valuable territory as possible.

If Gray had been a thorn in the side for the Mexican boundary commissioners, Major William H. Emory, another surveyor-astronomer, was a sword. He too repudiated the García Condé-Bartlett compromise. Emory, an original member of the U.S. commission, resigned in 1851, returned as surveyor and chief astronomer in 1853, then became the head commissioner in 1854. He disdained Mexicans of mixed blood, viewing them as examples of racial "decline."[16] His resolve to gain maximum American advantage in the boundary give-and-take is reflected in a statement he made expressing hope that the United States "might succeed in torturing the Treaty of Guadalupe-Hidalgo to embrace a practicable [railroad] route."[17] To Mexican diplomat and historian Cesar Sepúlveda, Emory's words and actions revealed a streak of expansionist greed. Sepúlveda concludes that the "ambitious and unscrupulous" Emory was willing to use censurable methods to get what he wanted. When it appeared to Emory that it might not be possible to obtain the desired railroad route, he recommended that the territory be taken by force. Emory, stated Sepúlveda, was guilty of provoking confrontation by setting off a "spark" of rather minor importance, but one that could ignite another war.[18] Even Bartlett, a learned man who seemed more fair-minded than his colleagues, must have irritated the members of the Mexican Commission. His contempt for biracial people and scorn for the Catholic faith are well illustrated in his *Personal Narrative*.[19] Such pejorative attitudes possibly surfaced from time to time in Bartlett's demeanor during negotiations between the two commissions.

Showdown at La Mesilla

Between December 1850, when García Condé and Bartlett achieved their famous compromise, and the ratification of the Gadsden Treaty three and a half years later, the Mesilla strip remained disputed land. The presence of Mexican repatriates in the region would make resolution of the New Mexico-Chihuahua boundary particularly difficult. From 1848 to 1852 nearly two thousand people wishing to escape U.S. jurisdiction, or seeking refuge from harassment at the hands of abusive Anglo Americans, settled at La Mesilla. A large percentage had lived in New Mexico east of the Rio Grande, an area claimed by Texas between 1836 and 1850. In his *Personal Narrative*, Bartlett described the circumstances that drove many of these refugees into Mexican territory:

> *Immediately preceding, and after the war with Mexico, the Mexican population occupying the eastern bank of the Rio Grande in Texas and New Mexico were greatly annoyed by the encroachments of the Americans, and by their determined efforts to despoil them of their landed property. This was done by the latter either settling among them, or in some instances forcibly occupying their dwellings and cultivated spots. In most cases, however, it was done by putting "Texas head-rights" on their property. These head-rights were grants issued by the State of Texas, generally embracing 640 acres, or a mile square, though they sometimes covered very large tracts. They were issued to persons who had served in her wars... and also to original settlers....*
>
> *With these land certificates, or "head-rights," many Americans flocked to... the Rio Grande, and in repeated instances, located them on property which for a century had been in the quiet possession of the descendants of the old Spanish colonists. The latter, to avoid litigation, and sometimes in fear of their lives, abandoned their homes, and sought a refuge on the Mexican side of the river.*[20]

Bartlett wrote that when he and García Condé agreed to the 32° 22' north latitude compromise in April 1850, confirming that La Mesilla was in Mexico, the fears of local residents that they might wind up in the United States disappeared. "The whole population had determined to abandon the place if the boundary line had run south of the village, and thus placed them under the jurisdiction of New Mexico.... The event was celebrated by firing of cannon and a grand ball.... After this the population continued to increase."[21] Governor Angél Trias of Chihuahua stationed several hundred soldiers at La Mesilla, although they were withdrawn within a short time because of shortages of funds.[22]

Once the U.S. Congress rejected the García Condé-Bartlett line, newly-appointed New Mexico governor William Carr Lane claimed the Mesilla strip for the United States and threatened to occupy it with military force. He was emboldened by advice given to him earlier by New Mexico's territorial delegates in the U.S. Congress to invade La Mesilla.[23] Lane asserted that the disputed area had been taken illegally by the state of Chihuahua, and that the residents of the town wished to be annexed to the United States. Indeed, a portion of the population of La Mesilla preferred to be in U.S. territory, but it appears they were a small minority comprised largely of Anglo American settlers. Lane's claim that the Mesillanos preferred the United States sounded preposterous to Mexican officials. They pointed out that most people residing in the town had fled the United States a short time earlier and without question they wished to remain in Mexico.[24] Bartlett shared the Mexican view of the situation at La Mesilla, commenting that fewer than twenty Anglo Americans lived there.[25]

Lane failed to obtain direct U.S. government or military support for his aggressive plan to occupy the land in question, so he led a volunteer force to Doña Ana. There, on 13 March 1853, he proclaimed that he was taking possession of the Mesilla territory until the boundary was determined by the United States and the Mexican Republic. He clarified that he was taking this action on his own, without orders from the Cabinet in Washington. Governor Angél Trias of Chihuahua countered by warning Lane that he would "use the means unquestionably necessary for its [La Mesilla's] defence [sic] and conservation, in case it is attacked," adding that the "responsibility for the consequences" would rest with Lane. With the backing of Mexico City, Trias followed with a military buildup at Paso del Norte.[26]

Once again armed confrontation over disputed land seemed imminent between the United States and Mexico, but a determination to resolve the matter through negotiation eventually prevailed. Lane, whose impetuous actions had contributed significantly to bringing about the crisis, unintentionally helped matters by resigning the New Mexico governorship for reasons unrelated to the border controversy. His replacement, David Meriwether, had specific instructions from Washington to "abstain from taking forcible possession of the tract, even if on your arrival in New Mexico you find it held adversely to the claim of the U.S. by Mexico or the authorities of Chihuahua."[27]

Conflict at the New Mexico-Chihuahua frontier eventually gave way to hard bargaining in Mexico City with the arrival in 1853 of newly-appointed U.S. Minister James Gadsden. Following in the tradition of several American envoys before him, Gadsden tried to coerce the Mexican government into selling large portions of its northern territories. But, while agreeing to sign the Gadsden Treaty, or El Tratado de La Mesilla, President Antonio López de Santa Anna would only consent to sell the strip of land long identified by the Americans as necessary for a U.S. railroad route to the Pacific.[28]

Unfortunately for the people of La Mesilla, their town became part of the tract acquired by the United States in 1854, forcing many of them to pull up stakes and relocate a second time into Mexican territory. Those who remained on U.S. soil received guarantees of equal treatment and protection under the U.S. Constitution, just as other Mexicans annexed in 1848 had obtained through the Treaty of Guadalupe Hidalgo.[29] In the years that followed the signing of the boundary treaties, however, Mexican Americans in New Mexico and elsewhere witnessed continuous violations of their rights.[30]

Aftermath of Conflict

Identifying the exact location of the U.S.-Mexico boundary proved to be a prolonged, frustrating experience for the officials and technical personnel who conducted the fieldwork, and for the political leaders in the distant capitals of both countries. While the tracing of the line from the Pacific to the confluence of the Colorado and Gila rivers, and from Paso del Norte to the Gulf of Mexico spawned little disagreement, fixing the dividing line between New Mexico and Chihuahua became a major problem. The root of the controversy over La Mesilla lay in the errors in the Disturnell map and the imprecise language in the Treaty of Guadalupe Hidalgo. The disparity in the two countries' strengths also contributed significantly to the outcome of the dispute. In the end the United States, by far the more powerful nation, acquired its desired land through the signing of the Gadsden Treaty.

From Mexico's point of view at the time of the boundary dispute, the main highlights are that it prevented a military invasion of the disputed zone and that it resisted another U.S. diplomatic offensive attempting to annex greater portions of the national domain. Mexican officials could also derive some satisfaction from having sold a remote and largely barren tract of land for $10 million. The loss of La Mesilla, however, represented another blow to national pride. Mexicans deeply resented the American repudiation of the Pedro García Condé-John Russell Bartlett compromise, the pressure tactics used by James Gadsden in Mexico City, and the changes made by the U.S. Senate to El Tratado de La Mesilla. The Mesilla episode added to an already strong legacy of Mexican resentment and suspicion of the United States.

NOTES

Surveying and Marking the U.S.-Mexico Boundary: The Mexican Perspective

1. See Cecil Robinson, ed. and trans., *The View from Chapultepec: Mexican Writers on the Mexican-American War* (Tucson: University of Arizona Press, 1989), for an excellent exposition of Mexican interpretations of events during the period, including views on American expansionism and racism.

2. Article V of the Treaty of Guadalupe Hidalgo specified that the boundary would proceed from the Gulf of Mexico "up the middle of that river [the Rio Grande, or Rio Bravo del Norte], following the deepest channel...." Charles I. Bevans, ed., *Treaties and Other International Agreements of the United States of America, 1776-1949* (Washington D.C.: State Department, 1972), Vol. 9: 794.

3. Ibid. The authority for determining the exact location of the boundary rested solely with the two commissions, each comprised of a commissioner and a surveyor.

4. Angela Moyano Pahissa, *México y Estados Unidos: Orígenes de una relación, 1819-1861* (Mexico, D.F.: Secretaría de Educación Pública, 1985), 176.

5. Luis G. Zorrilla, *Historia de las relaciones entre México y Los Estados Unidos de América, 1800-1958*, 2 vols. (México D.F.: Editorial Porrua, 1977), I: 240-241.

6. Ibid., 272, 496.

7. Moyano Pahissa, *México y Estados Unidos*, 177.

8. Leon C. Metz, *Border: The U.S.-Mexico Line* (El Paso: Mangan Books, 1989), 11-29.

9. Joseph Richard Werne, "Pedro García Condé: el trazado de limites con Estados Unidos desde el punto de vista mexicano 1848-1953," *Historia Mexicana*, vol. 36, no.1 (Julio-Septiembre, 1986), 121.

10. J. Fred Rippy, *The United States and Mexico* (New York: F.S. Crofts, 1926), 110.

11. Ibid., 110. The comment from García Condé's letter is quoted in Metz, *Border*, 47.

12. Werne, "Pedro García Condé," 125.

13. Metz, *Border*, 11, 13.

14. Zorrilla, *Historia de las relaciones*, I: 337.

15. Ibid., 336.

16. William H. Emory, *Report on the United States and Mexican Boundary Survey*, vol. I, Part I (Washington: 34th Congress, 1st Session, Senate Executive Document no. 108), 69.

17. Ibid., 51.

18. Cesar Sepúlveda, *La frontera de México: Historia, conflictos, 1762-1975* (Mexico, D.F.: Editorial Porrua, 1976), 74-75.

19. John Russell Bartlett, *Personal Narrative of Explorations and Incidents in Texas, New Mexico, California, Sonora, and Chihuahua*, 2 vols. (New York: Appleton Co., 1854), I: 39-40, 262, 427, 449, 484, 488-489; II: 299-300, 435.

20. Ibid., I: 212-214. Bartlett reports that more than half of the population of the town of Doña Ana moved to La Mesilla in 1851.

21. Ibid., 214-215.

22. Odie B. Faulk, *Too Far North ... Too Far South* (Los Angeles: Westernlore Press, 1967), 119.

23. Rippy, *The United States and Mexico*, 116.

24. Ibid., 115-116.

25. Bartlett, *Personal Narrative*, I: 215.

26. Rippy, *The United States and Mexico*, 117.

27. Ibid., 119.

28. For the Mexican perspective on the Gadsden or La Mesilla Treaty see Moyano Pahissa, *México y Estados Unidos*, 242-256, and Zorrilla, *Historia de las relaciones entre México y Los Estados Unidos de América*, vol. I: 335-360. The U.S. view is presented in Paul Neff Garber, *The Gadsden Treaty* (Gloucester, MA: Peter Smith, 1959).

29. Articles VIII and IX of the Treaty of Guadalupe Hidalgo specify the property, civil, and religious rights accorded Mexicans absorbed into the United States in 1848. Article V of the Gadsden Treaty reaffirms those rights for the residents of the Mesilla strip. Bevans, ed., *Treaties and Other International Agreements*, vol. 9: 796-797, 815.

30. See Rodolfo Acuña, *Occupied America: A History of Chicanos*, second edition, (New York: Harper and Row, 1981); Matt S. Meier and Feliciano Ribera, *Mexican Americans/American Mexicans: From Conquistadors to Chicanos* (New York: Hill and Wang, 1993). Moyano Pahissa, *México y Estados Unidos*, 173-241, discusses many of the violations of the Treaty of Guadalupe Hidalgo by the United States and their disastrous effect on Mexico and Mexican Americans.

GRAY SWEENEY

DRAWING BORDERS

Art and the Cultural Politics of the U.S.-Mexico Boundary Survey 1850-1853

Led by an unseen hand, the landscape-painters of America have contributed to further [a] *grand result; and never before has there been a nobler opportunity afforded the artist to aid in the growth of his native land, and to feel that, while ministering to his own love of the sublime and the beautiful, he was at the same time a teacher and a co-worker with the pioneer, the man of science, and the soldier, who cleared, surveyed and held this mighty continent, and brought it under the mild sway of civilization.*

— S. G. W. Benjamin, 1882

Decried as a government boondoggle, the survey of the Mexican boundary by John Russell Bartlett from 1850 to 1853 nearly led to a second war with the Republic of Mexico over the Mesilla Valley in southern New Mexico. However, the art and literature resulting from the expedition was of great significance. Through the publication of Bartlett's illustrated *Personal Narrative,* and the landscape paintings of Henry Cheever Pratt, an artist of the Hudson River School who accompanied Bartlett and became his collaborator, the alien yet wondrous desert landscapes of New Mexico and the Southwest were extensively depicted for easterners. Their images of the region created cultural foundations for an appreciation, and the eventual incorporation, of this vast and ecologically complex desert landscape. The U.S.-Mexico Boundary Survey, for all its alleged mishaps, was a key event in the historical process of constructing a visual and literary definition of the Southwest for mid-19th century Americans.

The U.S.-Mexico Boundary Survey was authorized at the highest levels of political power—Congress and the President. John R. Bartlett, a learned bibliophile and ethnographer, accomplished amateur artist, and Whig political appointee, was charged with drawing the official border between the United States and Mexico—a boundary running more than a thousand miles from El Paso on the Rio Grande to Fort Yuma on the Colorado River and then on to San Diego on the Pacific Ocean. Bartlett had no previous experience or training in such matters, but after a series of more experienced men, including famed explorer John C. Frémont, accepted and then turned down the position, Bartlett was appointed U. S. Commissioner. The year was 1850, and John Russell Bartlett would spend the next three years exploring the deserts of what are now the states of Texas, New Mexico, Arizona, and California, and the Mexican states of Chihuahua and Sonora.[1]

The Problems of Inscribing a Border on a Frontier

Bartlett's and Pratt's expedition, commissioned to draw a boundary across the desert, was necessitated by the Treaty of Guadalupe-Hidalgo that concluded the Mexican War. The U. S. victory stripped Mexico of almost half of its territory, taking what are now the states of Texas, New Mexico, Arizona, and California. According to the ideology of Manifest Destiny, expansion on such an imperial scale was exonerated by quasi-religious notions that the nation was divinely ordained to build a transcontinental empire from this military conquest. The treaty concluding the war stipulated that the final determination of the actual border between the United States and Mexico would be left to joint boundary commissioners and surveyors, working together. They would run the boundary line, and become, upon agreement, final signatories concluding the treaty.

Sectional interests and tensions revolving around the burning issue of the expansion of the slave system permeated American society, and also played a major role in complicating Bartlett's survey of the Mexican Boundary. John R. Bartlett, an anti-slavery Whig from Rhode Island, had little interest in advancing the agenda of expansionist Democrats and Southerners. Bartlett was unwilling to press the Mexicans to concede additional territory for a railroad route below the line he agreed to with the Mexican Commissioner at 32° 22' north latitude. His respect for the Mexicans, and his Yankee dislike of slavery, outweighed the insistence of southern politicians and their allies in his survey party that he seek further territorial gains. When the new Democratic administration of Millard Fillmore took office, Bartlett's enemies in Congress sought his dismissal. In the meantime, James Gadsden was quickly dispatched to negotiate the Gadsden Purchase, obtaining for the United States what one Texas senator called the "best route" for a railroad to the Pacific.[2] The same southern congressmen and senators who criticized Bartlett's boundary line were undoubtedly aware that the state of Texas had chartered the Texas Western Railroad in 1852 to build a railroad along the 32nd parallel—just south of the line Bartlett had agreed to with the Mexicans. Such a route could eventually link Charleston,

South Carolina, with San Diego and would give Southern interests a distinct advantage in colonizing the Southwest, further spreading the regime of slavery. The dispute about this narrow strip of land became so heated that armed conflict broke out in southern New Mexico between covetous American settlers and Mexicans who were being driven from their lands in the fertile Mesilla Valley—territory that Bartlett's "mistaken" boundary had returned to Mexico.

As if these political problems were not formidable enough, at the core of Bartlett's mission was a cultural paradox. He was supposed to mark a border in what was perceived in the East to be a raw, unsettled frontier. The Southwest was so remote and untouched by the cultural and market centers of the East that western history scholar Howard Lamar characterized the region Bartlett entered as a "feudal frontier."[3] The violent social conditions prevailing in this remote region seemed almost incomprehensible to the sedentary New England scholar. This was coupled with the alien appearance of the desert landscape, which seemed like a foreign land. The oppressive heat, parching aridity, and rugged topography of the desert Southwest did not conform well to the aesthetic conventions of landscape appreciation as then understood by educated Easterners. Yet it was exactly the strangeness of the Sonoran Desert, particularly the giant saguaro cactus, and the ancient aboriginal and later Hispanic cultures of the region that made it a fascinating territory for the learned Bartlett, while the sublime landscapes and Spanish Colonial architecture of the region were subjects of great interest to the landscape painter Pratt.

Besides their discovery of the place, the artist-explorers invented ways to represent this peculiar landscape for emigrants and for eastern investors hungry for information about the region. A major part of their work was the production of images that would make the desert intellectually and visually comprehensible to their audiences. Inscribed in Bartlett's and Pratt's paintings were complex sets of cultural signifiers whose associations and meaning may not always be obvious today.

To his critics, the most troublesome aspect of Bartlett's survey was that his intense intellectual curiosity frequently diverted him from his primary mission to mark a boundary. These diversions led him to explore places far beyond his official commission. Bartlett's travels took him to northern California's gold fields and its rich Napa Valley, and south into Sonora and Chihuahua. Intense criticism of his expedition, by his arch political enemy, army officer and Southerner William H. Emory, complicated matters. As the Official Astronomer for the expedition Emory opposed Bartlett's agreement with the Mexicans. Emory's criticisms of Bartlett kept the debate about Bartlett's survey alive well into the late 1850s.

Distracting Bartlett further from his mission were his duties as the U.S. government representative of law and order in this violent, unsettled region. When a member of Bartlett's party, the son of a senator from Rhode Island, was killed in an El Paso dance hall brawl, Bartlett enforced the law by catching, trying, and hanging the murderers. In another telling incident Bartlett took it upon himself to rescue a fifteen-year-old Mexican girl, Inez Gonzales, from being sold as a slave.

Bartlett was also surveying across territory claimed by the Apache nation that had nearly prevailed in a long guerrilla war against settlers. After centuries of resisting Spanish colonizers and Mexican residents, the Apaches took renewed offense when the United States began expanding across their lands, bringing new waves of immigrants. Settlements on this frontier, as Bartlett found, were little more than "an attenuated island chain awash in a sea of hostiles."[4] Although Bartlett and his party only once came under direct attack, they were keenly aware that the Apaches did not welcome them. When Bartlett found peaceful Pimas and Maricopas farming along the fertile Gila and Salt Rivers they came to represent the native peoples he could advocate to readers of his book.

Finally, Bartlett's suitability for the position of Commissioner was questioned. As the *New York Evening Post* stated bluntly, "Mr. Bartlett is a good botanist, and has some other knowledge, but [he] has no knowledge of camp life, and cannot properly control such a party as he has under him."[5] Leading a party of over one hundred men, with the difficulties of logistical support for personnel and animals in a wilderness frontier, was not Bartlett's forte. The real value in his work was in the cultural and historical record he produced.

For political reasons the government never published Bartlett's findings, like it did the reports of other official expeditions. Yet it is the very diversity of Bartlett's interests and his relative openness to the region's difference that continue to make his *Personal Narrative of Explorations and Incidents in Texas, New Mexico, California, Sonora, and Chihuahua, 1850-1853*, a key document of Southwestern history and art history. Bartlett's considerable talent as a literary stylist gave his book a wide and important readership. The book's illustrations of the exotic desert, along with the exhibitions of Pratt's paintings, were the first representations of the Southwest seen in the East.

Constructing a Visual Order for the Desert

After more than a decade of revisionist interpretations of American art, art historians have now come to understand that the images produced by artist-explorers like Bartlett and Pratt were more than mere documentary, or topographical views of landscapes as they "really looked."[6] Their work simultaneously documented a particular place or moment, chronicled the disappearing past, and promoted a vision of progress.[7] The visual images produced during the boundary survey are important because they reveal the varied ways that dominant ideologies can assimilate new phenomena, even those as bizarre and seemingly useless as the "barren" deserts of the Southwest, and encode them with a variety of iconological and ideological metaphors.

Historically, it is remarkable the way these images supported the commercial development of the region. By objectifying the Southwest for the first time, Bartlett and Pratt paved the way for its eventual incorporation by cultural and market centers in the East.[8] Developers and speculators realized that railroads would soon make these remote places accessible to settlers, to commerce, and to tourists, and they yearned for the economic development this would bring.[9] The fashionable appreciation of natural scenery, and a romantic nostalgia for native peoples, regional history, and exotic places are clearly evident in their illustrations and paintings.[10] Such images were powerful because in the popular mind these things were destined to pass away before the advance of American progress.

Bartlett and Pratt articulated their vision of the Southwest within a well-defined cultural perspective. Their achievement was to construct for a cultural elite in the eastern United States sets of associations that made this strange and forbidding landscape comprehensible as history and scenic wonder and commodifiable as national destiny.

The Cultural Work of the Artist-Explorers

Bartlett's survey of the U.S.-Mexico Boundary happened at a moment of extraordinary opportunity. Following its victory in the Mexican War, the federal government was eager to fund expeditions to its newly acquired territories in the Southwest. Expeditions of the 1840s, such as those of John C. Frémont and William H. Emory during the Mexican War, offered explorers and artists rare opportunities for government patronage, at a time when financial support was difficult to obtain.[11] These official expeditions were customarily followed by the publication of illustrated, multi-volume studies covering everything from topography and geology to botany, ethnology, and natural history. The publications provided maps and directions for emigrants, and often promoted routes for a long-sought transcontinental railroad, but they were very costly. One economic measure of the official importance attached to the boundary survey is the estimate that it consumed more than $750,000 during its entire duration. In 1853, the last year of Bartlett's expedition, the congressional appropriation for the survey was $200,000, an astonishing 3.5% of the Department of Interior budget.[12]

Bartlett's survey also occurred at a time when expansionist endeavors created support for the visual arts in the antebellum era.[13] Despite this great promise of government patronage for the eventual publication of the expedition reports, Bartlett and Pratt were to be bitterly disappointed. When Bartlett's account finally appeared, printed by a commercial publisher, there was an acerbic irony in his decision to title it a "*Personal Narrative*." The modest size of the book compelled Bartlett to eliminate most of Pratt's drawings, and the two volumes contained no expensive color reproductions.

Although he may not have been expert in field logistics, Bartlett was well qualified for the Boundary Commissioner's position intellectually, artistically, and politically. He associated himself with numerous learned societies in Providence, Rhode Island, and later in New York, where he became a leading book dealer. Born into a middle class family and raised an Episcopalian, Bartlett grew up in Kingston, Canada, where he first acquired his taste for history and good books. Bartlett was a close associate of Albert Gallatin, the Secretary of the Treasury under Presidents Jefferson and Madison. The two men founded the American Ethnological Society in 1842, and Bartlett later wrote a memoir of Gallatin. Bartlett was also friendly with John L. Stephens and Ephraim G. Squier, who he encouraged to visit Mexico and Central America, where they discovered the ancient civilizations of the Maya. Perhaps Bartlett harbored hopes he too might rediscover "lost" civilizations in the Southwest, as his colleagues had found in the Yucatán Peninsula.

A respected scholar, by 1847 Bartlett had published his *Progress of Ethnology*, and the next year his *Dictionary of Americanisms*, which went through several editions. Bartlett, as a bibliophile and importer of foreign books, was especially conversant with the Spanish Colonial literature on the Southwest. His *Personal Narrative* is peppered with extensive citations from 17th century sources such as Sebastián Viscaíno's *Voyages*, and Father Kino's writings. He also knew of the 18th century missionary writers Father Junipero Serra, from Alegro's history of the Jesuits in New Spain, and from Pedro Font's *Journal*. Bartlett was also familiar with modern literature on the region, particularly Baron Von Humboldt, to whom he often referred. He also knew the report of army officer William H. Emory, whose military reconnaissance across the region during the Mexican War preceded him. Because of these historical and literary interests it is not surprising that the last thirty years of Bartlett's life were spent in close association with John Carter Brown's library, where he helped to enlarge one of the great collections of early Americana.

Bartlett was also an amateur artist of considerable ability, as demonstrated by the more than two hundred images he returned East with. While he may have missed some of the subtle points of technique and the expression of "sentiment" that distinguished painters working in the best tradition of New York's National Academy of Design, or Boston's Athenaeum, the few oil paintings by Bartlett that have been located possess a solid technical facility and a thorough understanding of conventions of representation as practiced by artists of the Hudson River School. Bartlett's interest in art emerged during his early years in Providence, when he was working for the Globe Bank and helping to organize the Providence Athenaeum. At that time he decorated a cabinet of Egyptian antiquities and was zealous in his devotion to art. His early aspirations to art are evident in his submission in 1829 of three drawings for an exhibition in Providence.[14] In his "Autobiography," Bartlett later wrote that "nearly all the pictures now in my house painted by me were the work of my leisure hours while connected with this Bank."[15]

Bartlett knew Henry Cheever Pratt at least as early as 1830. In Pratt's "Catalogue of Portraits done at Providence R.I. 1830" at the Museum of Fine Arts, Boston, he lists a portrait of "Mr. John R. Bartlett" with a price of $15. Interestingly, among the portraits inventoried in his catalog, this portrait of Bartlett is not marked "paid." Perhaps, even at this early date, Bartlett established a pattern of promising advancement and opportunities to Pratt, but not cash, that was to repeat itself twenty years later. When Pratt painted Bartlett in El Paso during the time of the survey, he depicted him as an artist in a room filled with fine paintings. Bartlett eventually came to own several other oil paintings by Pratt, as well as retaining possession of most of the drawings Pratt made while in the Southwest.

Bartlett arrived in New York in 1836, remaining there through the period when artists like Thomas Cole and Asher B. Durand were establishing a new national identity for American landscape painting. Bartlett did not leave New York for Providence until 1849, so, although there is no documentation, Bartlett was undoubtedly aware of the connection between Pratt and the famous Cole. Historian Robert Hine speculated that through his extensive connections in New York's Whig elite, Cole was probably a visitor at Bartlett's well-known book store.[16] Bartlett's personal taste as a dilettante landscapist, and his keen awareness of the growing importance of landscape painting, may explain why he selected Pratt to accompany him to the faraway Southwest. Pratt thoroughly reflected some of the concerns of America's leading landscape painter, Cole.

Despite his reputation as a member of the New York literary establishment, Bartlett found it difficult to support his family, and in 1849 returned to Providence to begin searching for new financial opportunities. He had impeccable Whig credentials and friends in the highest political circles. With the support of Gallatin he gained an introduction to the powerful Washington senators Stephen A. Douglas, Thomas Hart Benton, and John C. Calhoun. They in turn introduced him to Jefferson Davis. These influential contacts enabled Bartlett to gain the boundary survey Commissioner's Office. His strongest advocate was Rhode Island Senator John H. Clarke, whose son would accompany Bartlett west. Bartlett may have also had the support of Washington's influential scientists Joseph Henry and Henry R. Schoolcraft.

However, Southerners worried about Bartlett's New England sympathies, and recalled that he had sided with Gallatin against the Mexican War. Some Democrat politicians even asserted that Bartlett was an abolitionist and a protégé of William Seward. Although Bartlett might have preferred an appointment in the foreign service, when the position of Boundary Commissioner, with its $3000 a year salary, was offered, he quickly accepted. In his unpublished autobiography, Bartlett revealed both a lack of experience and that his motives encompassed more than the engineering goals of the survey:

> *Although my life and pursuits have always been of a sedentary character I always had a great desire for travel, and particularly for exploring unknown regions. I had also ever felt a deep*

interest in the Indians, and was glad of an opportunity to be thrown among the wild tribes of the interior. I saw, too, that there would be a wide field for new exploration, and that if the government would permit these, I would prefer the office of Commissioner to that of any other.[17]

On 19 June 1850, Bartlett became the fourth man appointed Commissioner of the U.S.-Mexico Boundary Survey, and the second to act. Newspapers in Boston, New York, and Washington were enthusiastic. The New York *Courier* extolled his virtue and integrity and praised his scientific reputation.

John Russell Bartlett was an ideal representative of elite, established Northeastern Whig political ideals—an exemplar of what was then regarded as the "refined and cultivated" class. Bartlett and his handpicked artist, Pratt, were thus exceptionally well-positioned to interpret the Southwest in cultural terms. They would define the region's landscape scenery in ways that would avoid the "meager utilitarianism" and "sordid tendencies of modern civilization."[18]

While so much is known about the prolific and cosmopolitan scholar Bartlett, less is known about Henry Cheever Pratt. It was no surprise that Pratt was eager to accompany Bartlett's well-funded survey in the summer of 1851 when Bartlett offered him the position after another artist had been dismissed; a government salary was certainly welcome. Just a few years earlier Pratt had been lauded in the *Boston Evening Transcript* as a deserving artist "of high merit, to whom the public have not yet extended the full meed of approbation, regard, and what they most need, employment."[19] The opportunity to go west at government expense gave Pratt the prospect of financial security that had long eluded him. The expectation of having his work published in an illustrated, official government report was an added incentive. Pratt thought that a series of letters from the region would heighten public interest in his explorations, and that there might be profits from pictures of the Southwest that he would produce upon his return to Boston.

Henry Cheever Pratt was an artist decisively influenced by the more famous Cole. Pratt, born in 1803, was contemporary to the first generation of landscape painters that included Cole and Durand. Like his mentor Cole, Pratt was "discovered" by the artist, and inventor of the telegraph, Samuel F. B. Morse in New Hampshire around 1817.[20] After apprenticing with Morse, whom he helped to complete his famous painting of Congress, Pratt established himself as a portrait artist in Boston. Among his early commissions was a

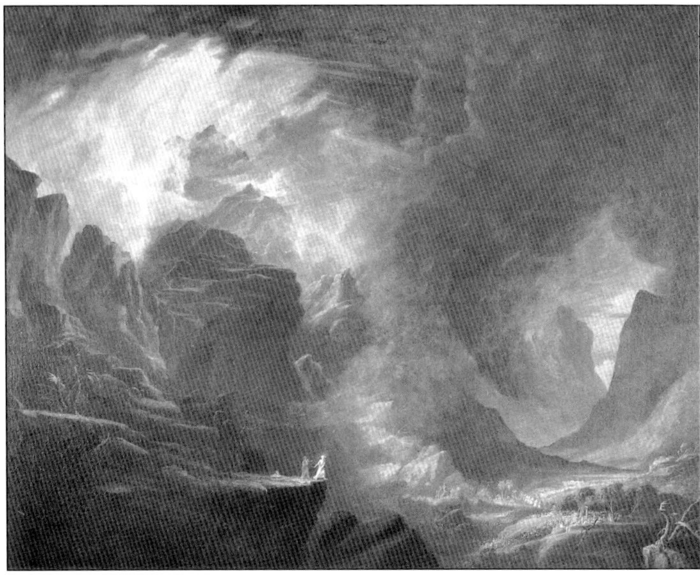

Figure 3
Henry Cheever Pratt
Moses on the Mount, ca. 1828-1829
46" x 60" • Oil on canvas
Shelburne Museum, Shelburne, Vermont

This painting was formally attributed to Thomas Cole.

portrait of Albert Gallatin. This commission again suggests the influence of his friend Bartlett.

Pratt's first major breakthrough in the emerging field of landscape painting came in 1828 when he accompanied Thomas Cole on a tour of the White Mountains of New Hampshire. The two artists set off, according to William Dunlap "sedulously studying the sublime of nature." They were guided by an itinerary prepared by Cole's patron Daniel Wadsworth.[21] The friendship between Cole and Pratt grew, and in 1844 Cole invited Pratt to join him on a trip to Maine's remote Mount Desert Island where they spent time sketching the scenery. One of Pratt's first paintings rendered in Cole's manner is the allegorical work *Moses on the Mount*, (Fig 3) which was once thought to be by Cole.[22]

In 1837 Pratt was trying to capitalize on Cole's notions that art should serve social authority, particularly the interests of clergymen. That year he exhibited his *Miracle of the Loaves and Fishes* at the Boston Mechanic's Association.[23] By 1848, Pratt, like his friend Cole, still did not realize that clergy and critics might favor allegorical and historical works, but patrons preferred to purchase portraits and landscapes. That year he advertised in the *Boston Evening Transcript* soliciting visitors to view his *Panorama of Paradise* (unlocated), a view of the Garden of Eden inspired by Milton, and modeled after the allegorical works of Cole.

By the mid-1840s, inspired by the success enjoyed by Cole, Pratt was confident enough to broaden his practice. He continued to produce portraits while exhibiting more landscapes, historical, and allegorical subjects. His 1843 painting *Indians Bathing - A Scene in New England* (Pl. 1) shows how directly he modeled his landscapes on Cole's. Its presentation of Indians in an Arcadian landscape is indebted to Cole's wilderness landscape as a type of "Savage State." Pratt's landscape is pure mythical history intended to carry the viewer back to a time before westward expansion eliminated the Native Americans of New England. Like many of Cole's pictures, it was based on verses that extolled this period as a kind of golden age when, "The maidens sport, amid the waters wild, Nor fear the warwhoop's scream or cannon's roar."[24]

For Whigs like Bartlett and Pratt, art and literature were meant to serve the needs of an elite social hierarchy, reinforcing the values of patrons—leaders of the Northeastern Whig establishment. They believed culture and art ameliorated the political excesses and economic instability of the Jacksonian Democrats. In Bartlett's view, they might help undercut the imperial ambitions of southern Democrats to expand the slave system across the West. It was no coincidence that, in 1852, in the middle of Bartlett's survey, Congress passed the Wilmot Proviso, preventing the expansion of slavery into territories taken from Mexico.

THE OFFICIAL DRAUGHTSMAN AND THE BOUNDARY COMMISSIONER

As the official "draughtsman" for the U.S.-Mexico Boundary Survey, Henry Cheever Pratt's responsibility was to make sketches of the important scenery along the route. These would later be used to illustrate the planned official reports. In this objective he was accompanied, perhaps even surpassed, by Commissioner Bartlett. The Southwest emerged in the images produced by the two artists as a frontier region of ancient Indian cultures and ruined cities, replete with a legendary Spanish colonial past. Rather than depicting hostile Indians, Pratt sketched peaceful natives who were agriculturalists already converted to European ways. The landscape signified enormous mineral wealth and great scenic potential for these artist-explorers. Most importantly, the Southwest represented the future, a place where an American vision of progress would expand its immense enterprise under the ideology of Manifest Destiny.

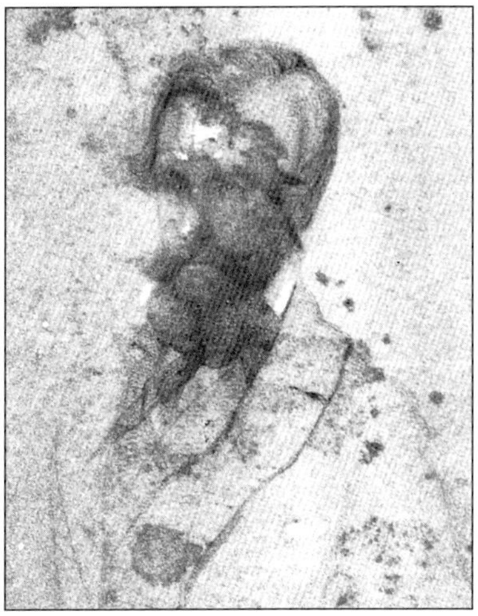

Figure 4
Augustus de Vaudricourt
Portrait of John Russell Bartlett (detail), ca. 1850
7 1/2" x 9" • Pencil
The John Carter Brown Library, Brown University, Providence, Rhode Island [Ø]

Pratt's predecessor, the French-born lithographer and art teacher Augustus de Vaudricourt, learned that being an artist on the boundary survey might have its financial rewards, but was also fraught with very real professional perils. Trekking through a wilderness was strenuous, but scarcely compared with the frustrations and pettiness inherent in a large-scale government project. Bartlett considered de Vaudricourt to be a worthless drunkard and eventually succeeded in having him dismissed. Pratt was invited to join Bartlett's survey party at this time. Vaudricourt's drawings made during his engagement on the survey are almost entirely unlocated. One exception is a sensitive pencil drawing of Bartlett (Fig. 4).

Soon after his arrival in El Paso in the summer of 1851, Pratt found that he was paid irregularly, so he began to make portraits to subsist. In a lengthy letter from Pratt to Bartlett written after his arrival on 23 July 1851, he pleaded, "I am getting good prices for my work here, and I am so much in need of the money, that I naturally wish to stay and earn what I can to so much advantage, I shall do double work for you when I come to make up for this."[25] Pratt described his prospects for patronage in El Paso, and his hopes that he would eventually receive commissions to paint the town's leading American traders James Wiley Magoffin and Benjamin Franklin Coons.

Bartlett's desire for self-promotion made it necessary for Pratt to paint an official Boundary Commissioner portrait, which he completed shortly before Bartlett disbanded his survey party and returned to the East (Pl. 2). This portrait had a retrospective aspect, recalling issues and events that had occurred over the course of the survey. The portrait was acquired by James Wiley Magoffin, El Paso's leading citizen. Magoffin may have looked on his acquisition as a memento of the hospitality he showed the Commissioner, and also as a gesture of repayment of Bartlett's great debt. The El Paso trader advanced the survey party much needed supplies and accepted Bartlett's government bank drafts even though they were not always backed by sufficient funds. The portrait is inscribed on the verso "Painted by Henry C. Pratt (U.S.B.Co) October 1, 1852." Pratt's portrait of Magoffin, which is identical in size to the portrait of Bartlett, is dated the next day, 2 October, 1852. Possession of such images by Magoffin conferred on him additional respectability as the town's leading citizen and reinforced his role as a self-appointed representative of culture and authority in the state.

Pratt's specialization was portraiture, and his depiction of Bartlett is a carefully constructed work designed to signify the Commissioner's status. During his years in Boston Pratt became expert at encoding into his portraits the key markers of his sitters' social status. His skills are evident in his portrait of Commissioner Bartlett, his long-time friend and now his official patron and superior. Pratt's *Portrait of John Russell Bartlett* is especially engaging because it shows the Commissioner primarily as a connoisseur of art. Painted in early October 1852, in the heat and rough frontier conditions of El Paso, the Boundary Commissioner is depicted formally in the portrait; he wears a gentleman's black waistcoat, starched white shirt, and cravat. Bartlett's intense gaze commands attention; his eyes are framed by his prominent forehead, and full, finely coiffured beard. Although Pratt's portrait of Bartlett represents him as a man of social substance and intellectual authority, it omits any reference to him as a scientist or bibliophile. Bartlett leans against a desk upon which are writing instruments and a clasp-bound ledger or sketch book, and on the brass support behind him are two large maps. One of these may have been intended to be a veiled reference to the Disturnell Treaty Map that had proved so troublesome to Bartlett's survey. Behind Bartlett's head the walls of the spacious room are covered with paintings in elaborate gilded frames. At the top left are two pictures of female subjects. One appears to be a half-nude female Indian, the other a woman with a red head scarf and blue robe. Although somewhat indistinct, the former may suggest Pratt's portraits of Maricopa-Pima Indian women, the latter may allude to his portrait of Inez Gonzales. (Both of these paintings are now unlocated, and these sketchy renderings may provide a clue to their appearance.) Below, in a lower tier, are two horizontal landscapes. One depicts mountainous scenery and may reference Pratt's now unlocated painting of the Organ Mountains. Two frames with prominent finials, placed just behind Bartlett's head, surround him with the cultural authority of high art. Whigs like Pratt advocated Cole's belief that art could enforce "good principles or the cause of virtue and morality."[26] The posed formality and status-conscious signification in the more highly mediated form of the oil on canvas contrasts strikingly with a previously unidentified pencil sketch of Bartlett by Pratt. Executed immediately and on the spot, it represents another side of the artist's patron: Bartlett is shown informally and at work standing before a writing desk (Fig. 5).

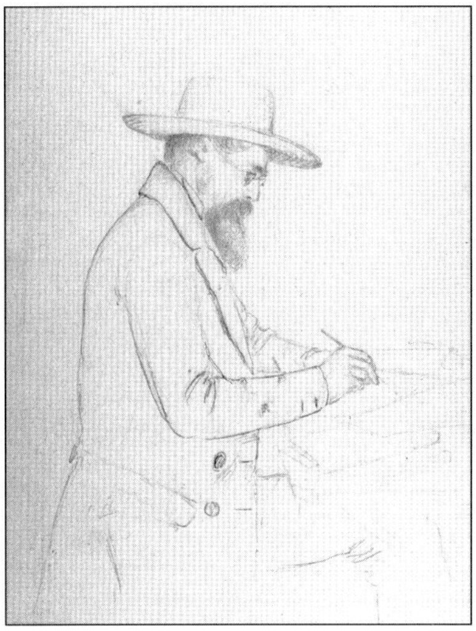

Figure 5
Henry Cheever Pratt
Portrait of John Russell Bartlett, ca. 1852
10" x 7 1/4" • Pencil
The John Carter Brown Library, Brown University, Providence, Rhode Island [166]

Self-Depictions of the Boundary Survey

Bartlett was intensely aware of the importance of his survey mission. As if to confirm this, he produced a number of images that self-represented his party as they explored across what he often called the "inhospitable wilderness."[27] Artist-explorers frequently felt the need to depict themselves, generally as small figures placed in the foregrounds of expansive landscapes, to visually authenticate and contextualize their passage through the region. Such diminutive figures were at one time routinely dismissed as mere "staffage," but art historian John Barrell argues that they should not be underestimated in establishing a cultural, and particularly a social signification in landscape paintings.[28] By including themselves in their images Bartlett and Pratt were self-referencing and self-promoting their mission as representatives of a triumphant power.

Very early in the expedition, while the party was crossing West Texas on its way to El Paso, Bartlett produced two self-depictions

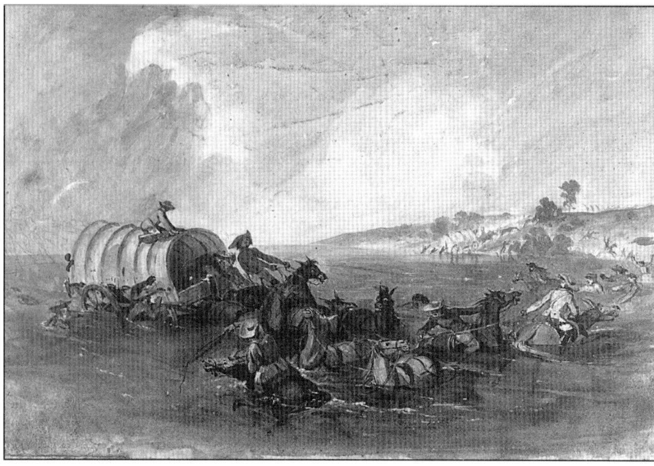

Figure 6
John Russell Bartlett
Crossing the Pecos, October 30, 1850
9" x 12" • Pencil and sepia wash
The John Carter Brown Library, Brown University, Providence, Rhode Island [64]

that illustrated the difficulties of the march. *Crossing the Pecos* (Fig. 6) dramatically shows the party fording the flooded Pecos River, a time when panicked mules nearly overturned the explorers' wagons. *Camp in Snow Storm on Delaware Creek, Texas* (Fig. 7) represents a bitter November snowstorm that halted the party. At Delaware Creek a foretaste of the hardships to come forced the survey team to burn wrecked wagons, abandoned along the trail by previous travelers, for warmth. Bartlett used the depictions of both incidents as the first full page illustrations of Volume I of the *Personal Narrative.* He then dedicated most of the commentary to the difficulties of the march—poor or non-existent roads, lack of water and feed for the animals, and the constant fear of Indian attacks. These were the kind of indispensable adventure narratives of danger and hardship expected by readers. The accompanying images authenticated this testimony, illustrating the adversities encountered in passing through such landscapes.

Bartlett avoided the rigors of camp life by living almost full time inside his elegant, specially built Rockaway carriage. It was a large vehicle with closed sides, curtained windows, and the ability to become a sleeper. The coach was dubbed a rolling armory because it sported so many firearms that Bartlett boasted he and his drivers could discharge thirty-seven rounds without reloading. Bartlett often inserted his carriage and the expedition's long wagon trains into his images as a means of self-referencing his presence. For example, in the forbidding landscape titled *Guadalupe Mountain, Texas* the great coach, and its escort, appears to move along in splendid isolation, shown almost as if it were riding on a paved road (Fig. 8). The sight of these mountains recalled to Bartlett familiar associations: "Often have I gazed at the Katskill [sic] Mountains in sailing down the Hudson; and though at a distance of but twelve miles, I never saw them as distinctly, as the Guadalupe Mountain appeared sixty miles off."[29] Upon closer approach to the sheer mountain pass, Bartlett invoked the standard rhetoric of landscapists. The view was so sublime that, "No sunrise at sea or from the mountain's summit could equal in grandeur that which we now beheld, when the first rays struck the snow-clad mountain, which

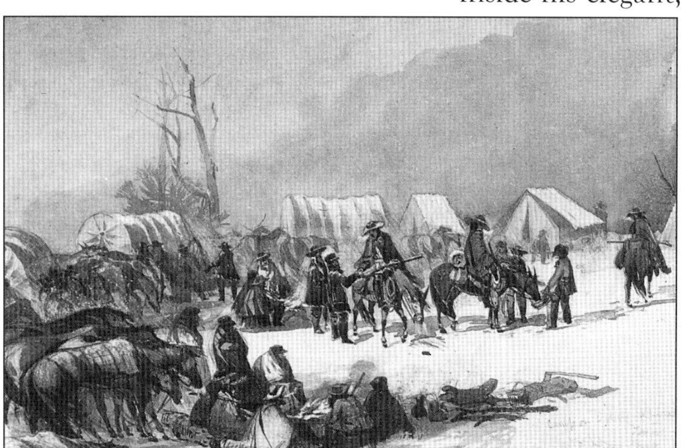

Figure 7
John Russell Bartlett
Camp in a Snow Storm on Delaware Creek, Texas, November 6, 1850
10 1/8" x 15 1/8" • Pencil and sepia wash
The John Carter Brown Library, Brown University, Providence, Rhode Island [34]

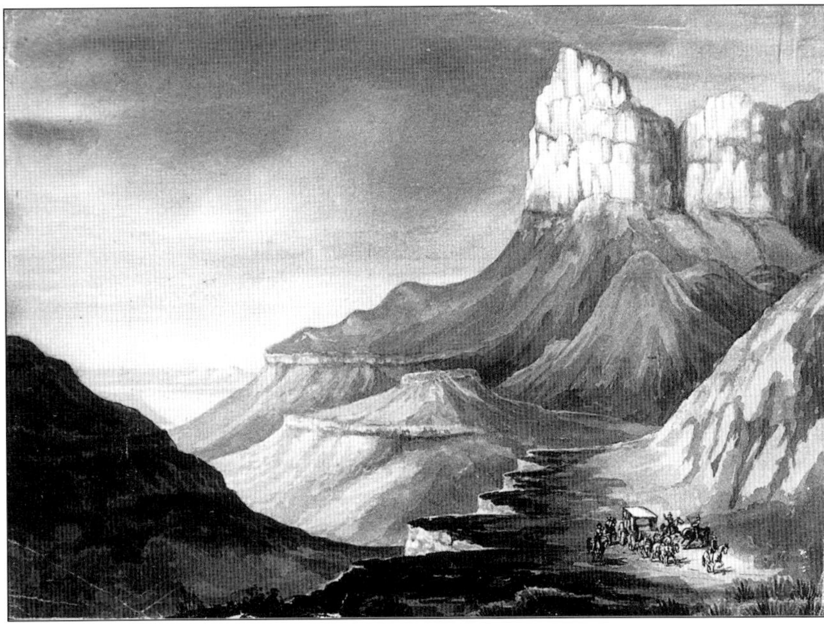

Figure 8
John Russell Bartlett
Guadalupe Mountain, Texas, November 10, 1850
9 1/2" x 13" • Pencil and sepia wash
The John Carter Brown Library, Brown University, Providence, Rhode Island [35]

reared its lofty head before us.... No painter's art could reproduce, or colors imitate, these gorgeous prismatic tints...."[30]

Prospective emigrants were an important audience for the *Personal Narrative.* Therefore, Bartlett often represented even the greatest difficulties of the march as strenuous but triumphantly overcome. To cross the vast landscapes, the best efforts of men and animals were called forth on the trek. The most telling example of this is Bartlett's drawing of *Guadalupe Pass, Cooke's Road, Sonora* (Fig. 9). Here, struggling Mexican laborers, pulling wagons up the steep pass in the foreground, are framed amidst the exotic vegetation and strangely anthropomorphic rocks in the distance. The curious desert vegetation and picturesque scenery of the pass starkly contrast the surrounding landscape. Bartlett called it "barren and uninteresting in the extreme." Confronting the harshness of the situation, doubts intruded in Bartlett's mind. In his text he confessed to the reality that is down played in his images:

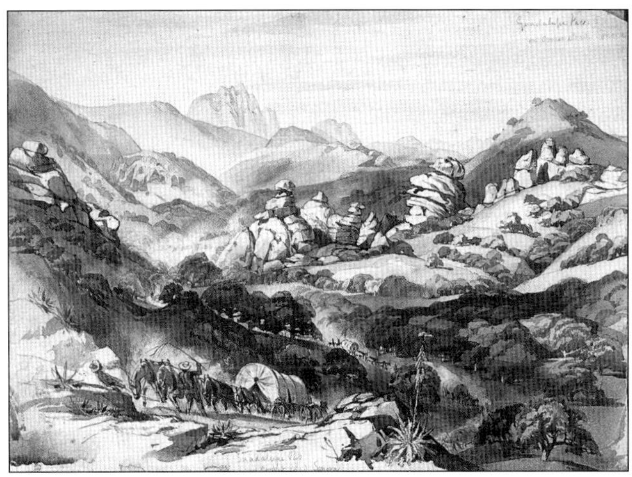

Figure 9
John Russell Bartlett
Guadalupe Pass, Cooke's Road, Sonora, ca. 1851
11 1/2" x 15 7/8" • Sepia wash
The John Carter Brown Library, Brown University, Providence, Rhode Island [41]

> *As we toiled across these sterile plains, where no tree offered its friendly shade, the sun glowing fiercely, and the wind hot from the parched earth, cracking the lips and burning the eyes, the thought would keep suggesting itself. Is this the land*

Figure 10
John Russell Bartlett
Fording a Stream, Packmules Sink in Quicksand, September 13, 1851
10" x 12 3/4" • Pencil and sepia wash
The John Carter Brown Library, Brown University, Providence, Rhode Island [117]

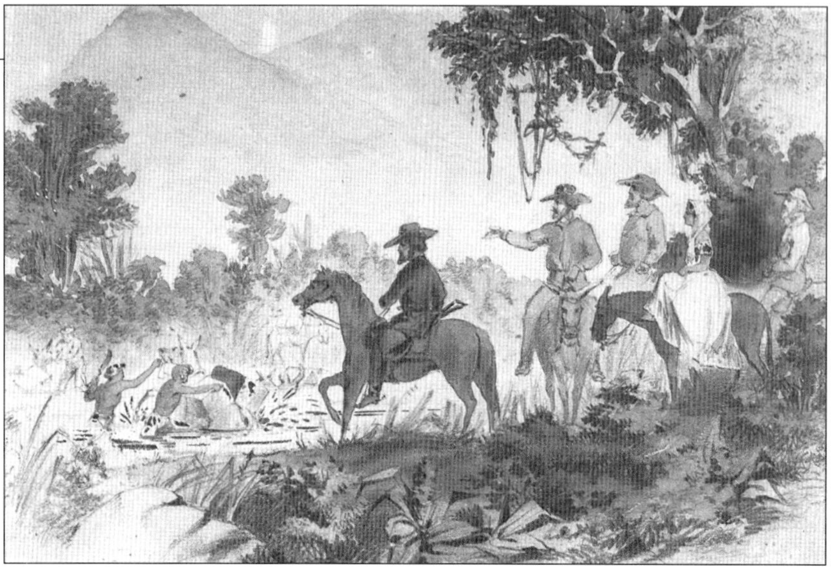

Fig. 10

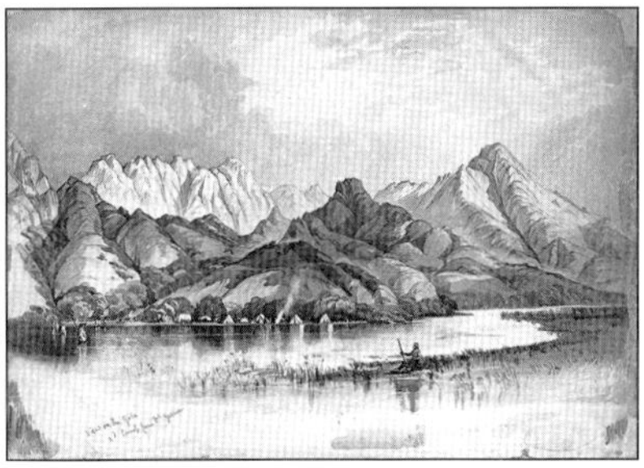

Figure 11
John Russell Bartlett
View on the Gila, 2d Camp from Ft. Yuma, June 1852
13" x 18" • Pencil and sepia wash
The John Carter Brown Library, Brown University, Providence, Rhode Island [95]

which we have purchased, and are to survey and keep at such as cost? As far as the eye can reach stretches one unbroken waste, barren, wild, and worthless. For fifty-two long miles we have traversed it without finding a drop of water....[31]

A similar objectification, and simultaneously an idealization of the difficulties of the march, was evident in the Commissioner's drawing, *Fording a Stream, Packmules Sink in Quicksand* (Fig. 10). Survey party *arrieros*, or muleteers, struggle to save drowning mules, while at the center a bearded man, probably Bartlett himself, gestures with noblesse oblige toward the scene almost as if he were giving a lecture at the Athenaeum.

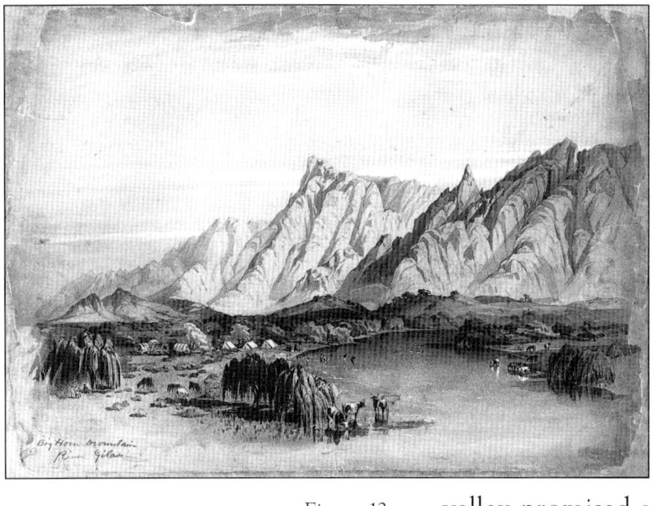

Figure 12
John Russell Bartlett
Big Horn Mountains, Rio Gila, June 22-23, 1852
12 3/4" x 18" • Pencil and sepia wash
The John Carter Brown Library, Brown University, Providence, Rhode Island [68]

At the other extreme were pastoral images of camp life. Bartlett's *View on the Gila, 2d Camp from Fort Yuma* and *Big Horn Mountains, Rio Gila* are representative (Figs. 11 & 12). Reaching the *terra incognita* around the Gila River was a prime objective of the expedition. Its river valley promised a potential route for a railroad, and its survey was one of the few specific objectives stated in Bartlett's official commission. In *View on the Gila River, Arizona* (Fig. 13) the explorers' camp nestles against the foot of the mountains, while in the foreground a wide river flows. On the right bank a friendly-looking Indian observes the party making camp. The figure in the middle of the river may be Pratt who was noted as something of a fisherman.[32]

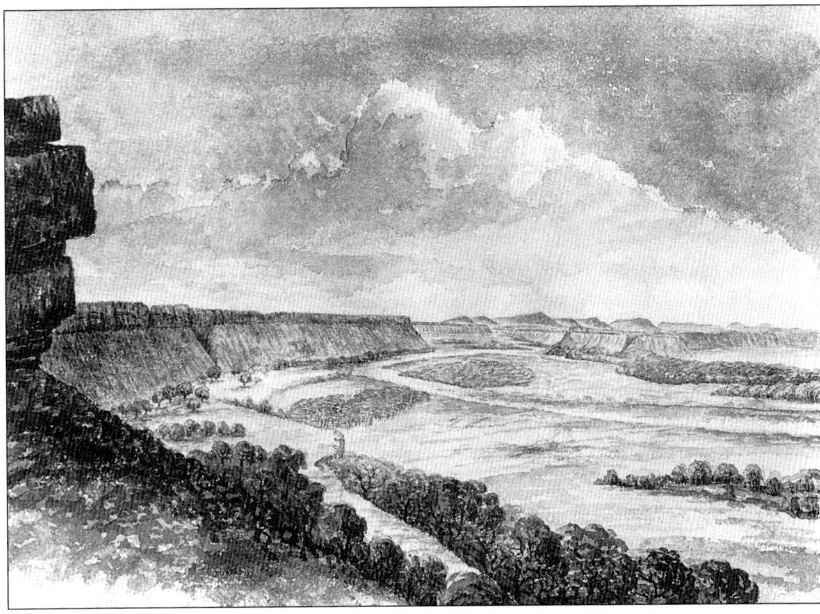

Figure 13
Henry Cheever Pratt
View on the Gila River, Arizona,
June 26, 1852
14" x 10" • Watercolor
The John Carter Brown Library, Brown University, Providence, Rhode Island [8]

Bartlett presented a similar visual construction in his *Big Horn Mountains, Rio Gila* (Fig. 12). In this drawing the rugged mountains in the background frame a pastoral camp scene in the foreground. Cattle drink from the river and graze amid the trees in the fertile river bottom, while the party bathes in the abundant water. Both scenes are inviting, suggesting that the well-watered Gila River Valley was suited to agriculture. Similar sentiments prevailed in several pencil drawings. In Bartlett's intricately executed pencil and wash drawing *On the Salinas, North of the Gila, New Mexico* (Fig. 14), his survey party is encamped peacefully with Indians at a site surrounded by trees overlooking the river. Other drawings such as *Camp at Small Lagoon on the Gila* (Fig. 15), and *An Encampment* (Fig. 16), reiterated the preference of the artist for landscapes with trees and water. Later, Pratt constructed several important oils that prominently depicted the surveying party as it moved through the landscape. In contrast to the images that supported Bartlett's ad-

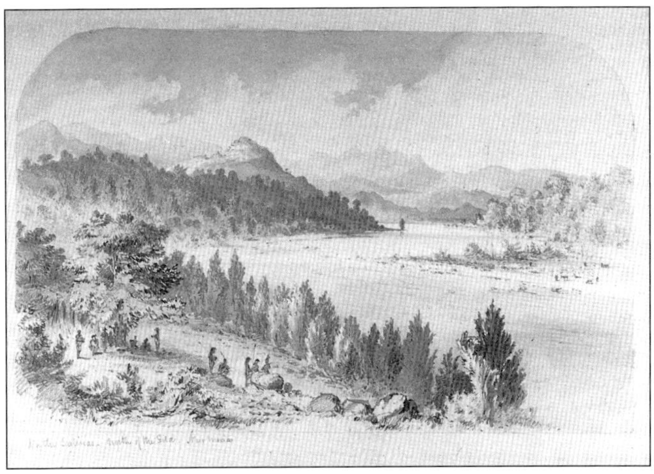

Figure 14
John Russell Bartlett
On the Salinas, North of the Gila, New Mexico, July 3-4, 1852
7 1/4" x 10 3/4" • Pencil
The John Carter Brown Library, Brown University, Providence, Rhode Island [98]

Figure 15
John Russell Bartlett
Camp at Small Lagoon on the Gila,
June 23 or 28, 1852
8 1/2" x 14" • Pencil
The John Carter Brown Library, Brown University, Providence, Rhode Island [84]

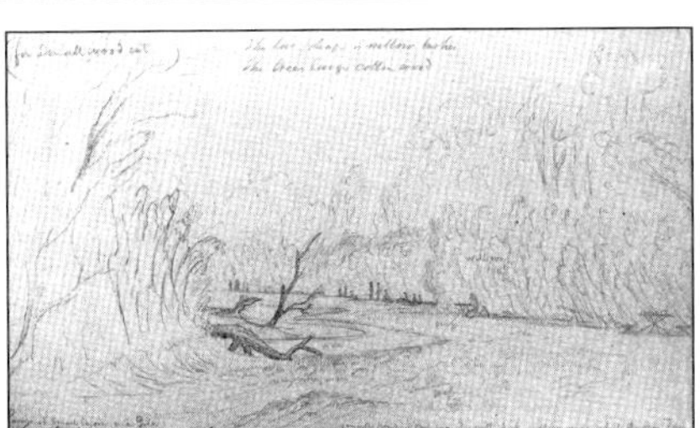

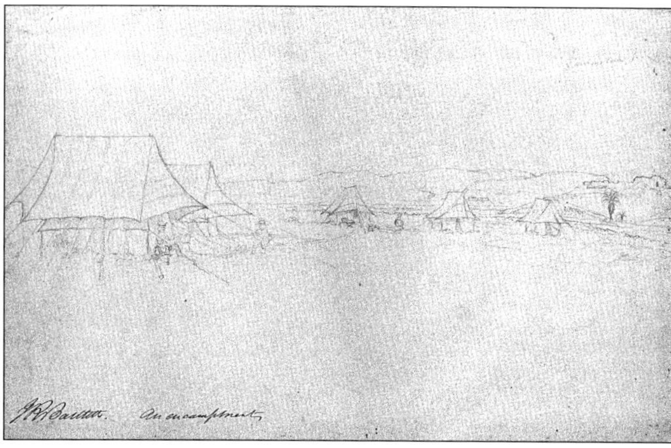

Figure 16
John Russell Bartlett
An Encampment, ca. 1852
9 1/8" x 14 1/4" • Pencil
The John Carter Brown Library, Brown University, Providence, Rhode Island [103]

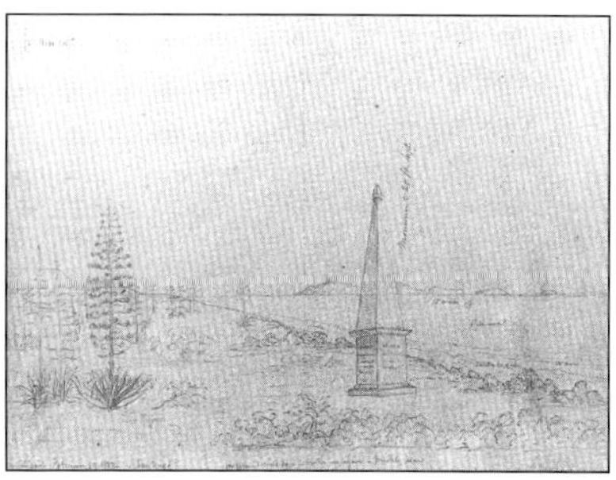

Figure 17
John Russell Bartlett
Initial Point, San Diego, February 19, 1852
9 3/4" x 13" • Pencil
The John Carter Brown Library, Brown University, Providence, Rhode Island [65]

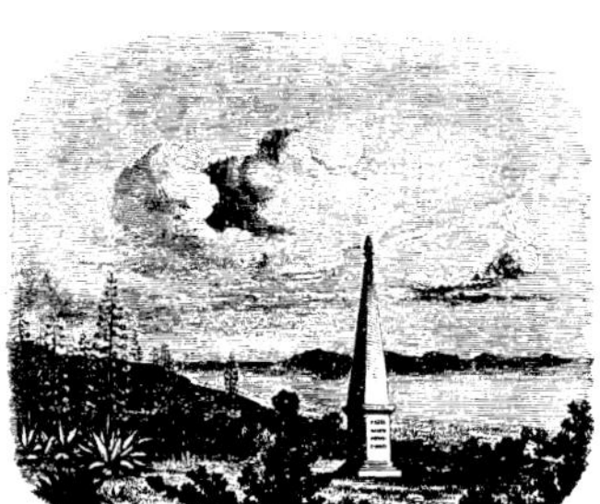

Figure 18
Monument at Initial Point, Pacific
Woodcut illustration from John Russell Bartlett's *Personal Narrative*

venture narratives, these images, like Bartlett's pastoral drawings, deemphasized the harsh desert conditions, idealizing the explorers' march through the wilderness.

The Work of the Boundary Survey

Only a handful of drawings directly related to the main work of the survey, marking a boundary line, were executed; this omission was a telling comment on Bartlett's real interests. Amid the hundreds of images from the expedition there are no depictions of surveyors marking the boundary line with transits, levels, barometers, or chains—an indication of Bartlett's evident disinterest in technical matters. The most important drawing related to the work of the survey is Bartlett's depiction of the elaborate architectural monument at the *Initial Point* in San Diego. With its numerous layers of symbolism, this monument marked the Pacific terminus of the boundary line (Fig. 17). Built by Bartlett's predecessor who had surveyed and marked the boundary line from San Diego as far east as Fort Yuma on the Colorado River, the imported obelisk of polished white marble, resting on a pedestal about twenty feet in height, marked the exact spot where the boundary line began. According to the treaty this was "one marine league south of the southernmost point of the Bay of San Diego." The *New York Evening Post* called it a "splendid marble monument," and gave details of its construction and inscriptions. The economic potential symbolized by the monument became evident in Bartlett's comments following his description of the obelisk. He exclaimed that "a finer agricultural country does not exist on the face of the globe."[33] In the woodcut reproduced in the *Personal Narrative* a halo-like cloud behind the obelisk functions to underscore the region's great promise (Fig. 18).

The eastern terminus of the boundary, marked by a simple stone cairn erected about forty miles north of El Paso at 32°22' north latitude, was the other major point along the survey route. With great ceremony Bartlett and the Mexican Boundary Commissioner Pedro García Condé signed documents that were placed inside the stone marker, including a fragment of the Washington Monument.[34] Bartlett's agreement with the Mexican delegation about this initial point later proved to be immensely controversial. No sketch of this simple marker is known.

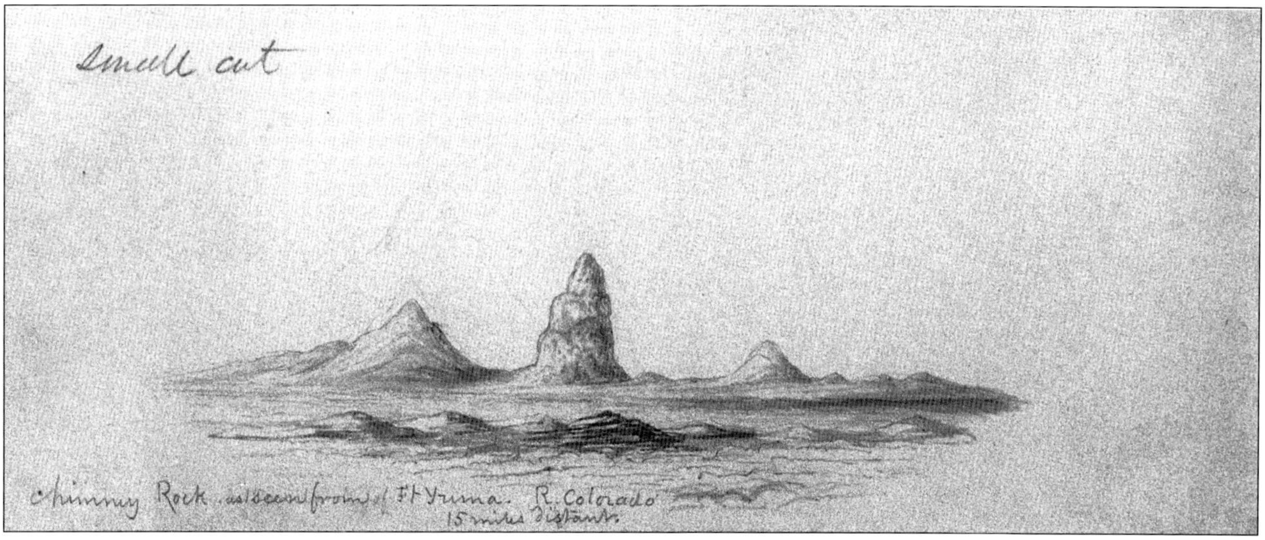

Figure 19
Henry Cheever Pratt
Chimney Rock as seen from Ft. Yuma, River, Colorado, June 16, 1852
3 3/4" x 9" • Pencil
The John Carter Brown Library, Brown University, Providence, Rhode Island [14]

The only other images directly related to the work of the survey are Pratt's impressive watercolor *Fort Yuma, Colorado River* (Pl. 3), and his sepia drawing *The Junction of the Colorado and Gila Rivers at Fort Yuma* (Pl. 4). The former image became the frontispiece for Volume I of the *Personal Narrative.* The latter was important because according to Article V of the Treaty of Guadalupe Hidalgo "a straight line shall be drawn from the middle of the Rio Gila, where it unites with the Colorado, to a point on the coast of the Pacific Ocean ... at the Port of San Diego." These drawings represented the western terminus of Bartlett's survey. Fort Yuma, the only ferry crossing on the Colorado River, was greatly feared as a wild and desolate area and was associated with Yuma Indian raids on emigrants bound for California. The landmark known as Chimney Rock is shown in the distance in Pratt's drawing. He also made a pencil sketch of this famous rock, *Chimney Rock as seen from Ft. Yuma, River, Colorado.* Because of these associations it was a marker long anticipated and equally dreaded by settlers moving westward (Fig. 19). The sight of Chimney Rock confirmed that they had succeeded in crossing the deserts of New Mexico, but faced one last obstacle, the great California Desert, before reaching their "promised land."

Frontier Life in Old El Paso

After reaching El Paso, Bartlett was confronted with a chaotic frontier reality. El Paso was to be his headquarters for much of the expedition, but the difficulties posed were almost unbearable. Provisions for his large party were exorbitantly priced, and the severe hardships of the route from Indianola on the coast of Texas to El Paso devastated his mules and horses, which also had to be replaced at great cost. These logistical and monetary problems paled, though, in comparison to the bands of "gamblers, horse thieves, and murderers" that had taken over the town. Bartlett described the situation: "Houses were opened for the indulgence of every wicked passion; and each midnight hour heralded new violent and often bloody scenes.... None dared stir from home without being doubly armed... for the turning of a corner might bring one to the muzzles of a dozen pistols."[35] The situation reached a climax following a "fandango," or community dance. Edward Clarke, the son of Rhode Island's Senator Clarke, who had strongly advocated Bartlett for the Commissioner's position, was brutally stabbed to death by a band of renegade Americans who were in the vicinity. An armed posse of officers and men from the survey party and the town soon located the murderers; they were brought to trial. The jury of six Mexicans and six Anglos quickly convicted the desperados. Bartlett's description of the proceedings gave a vivid portrait of the situation:

> *The trial took place in one of the adobe or mud-built houses peculiar to the country.... In the court room ... all were armed save the prisoners. There sat the judge, with a pistol lying on the table before him; the clerks and attorneys wore revolvers at their sides; and the jurors were ... armed, ... presenting a scene more characteristic of feudal times than of the nineteenth century.*[36]

Bartlett was keenly aware that Americans were only the most recent emigrants to the region; the old town of El Paso was founded in 1598. "Long before the consecration of Plymouth Rock, the religion of Christ had been made known to the Indians of New Mexico," Bartlett declared, "and the Gila and Colorado Rivers, which in our day are attracting so much interest as novelties, were passed again and again by the persevering and energetic Spaniard."[37] Potential for agriculture around El Paso particularly interested him. Both Bartlett and Pratt made drawings of the town and its surroundings. Bartlett was fascinated by the adobe buildings of El Paso, which to him recalled those of the ancient Assyrians. Bartlett produced a

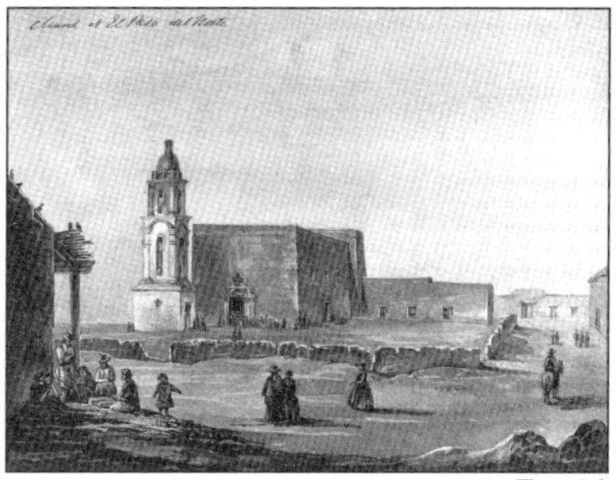

Fig. 20

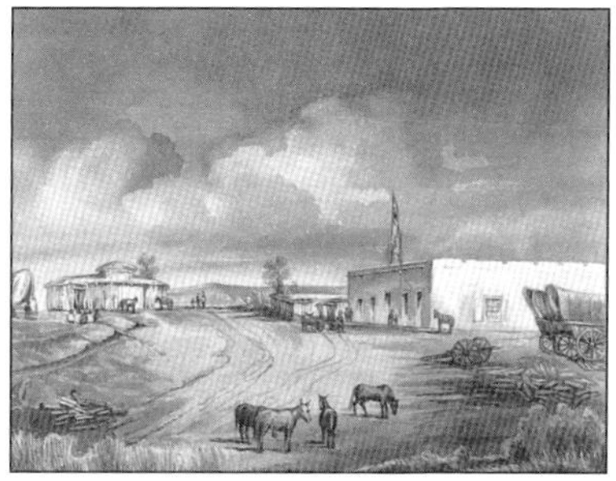

Fig. 21

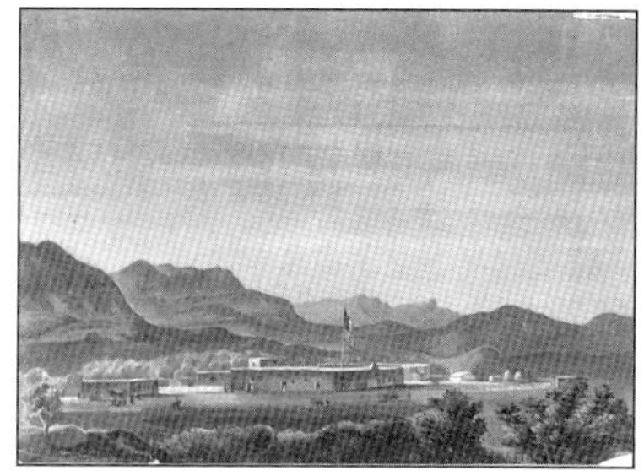

Fig. 22

Figure 20
John Russell Bartlett
Church at El Paso del Norte, Chihuahua, November 23, 1850
9" x 11 1/4" • Sepia wash
The John Carter Brown Library, Brown University, Providence, Rhode Island [3]

Figure 21
John Russell Bartlett
El Paso, Texas, ca. 1850-1851
8 7/8" x 11 1/4" • Sepia wash
The John Carter Brown Library, Brown University, Providence, Rhode Island [83]

Figure 22
John Russell Bartlett
Military Post, El Paso, Texas, August, 1852
8 1/8" x 11 3/8" • Sepia wash
The John Carter Brown Library, Brown University, Providence, Rhode Island [81]

Figure 23
Henry Cheever Pratt
Ruins of the Church at San Elizario on the Rio Grande, October 8, 1852
8 1/2" x 6" • Pencil
The John Carter Brown Library, Brown University, Providence, Rhode Island [17]

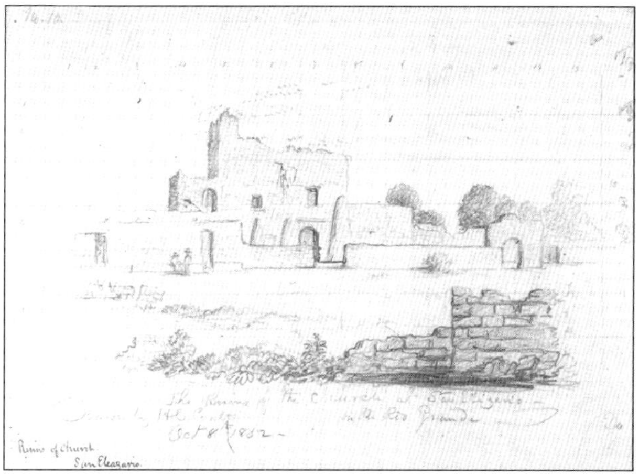

Fig. 23

drawing of the church and plaza at El Paso (Fig. 20), and also sketched James Magoffin's recent settlement (Fig. 21). "Its enterprising proprietor has erected around a large open square some of the best buildings in the country."[38] The original Magoffin house, visible in this drawing, was destroyed in the flood of 1868, but its successor, built in a similar style, is today a national historic landmark and a Texas State Park. Bartlett also mindfully made a drawing of the military outpost at El Paso, the symbol of American authority over the region, at Coons Rancho, a place Pratt would later paint (Fig. 22). At San Elizario, south of El Paso on the Rio Grande, Pratt sketched the ruins of the old Spanish mission church and Presidio that had served as the military outpost for American troops immediately following the Mexican War (Fig. 23). Pratt also made a panoramic sketch showing the substantial Mexican town of Paso del Norte (Fig. 24).

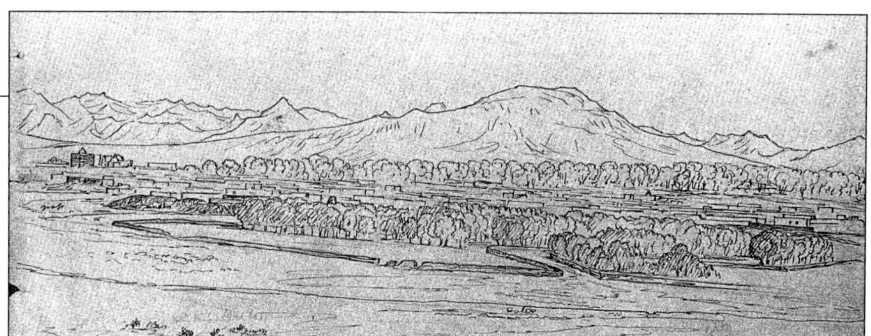

Figure 24
John Russell Bartlett
El Paso (detail), November 1850
9 3/4" x 14" • Pen and ink
The John Carter Brown Library, Brown University, Providence, Rhode Island [135]

Pratt convinced Magoffin to sit for his portrait at the same time he produced his portrait of Bartlett; while his portrait of Coons was completed in late July 1851. It is interesting to speculate whether Pratt fully grasped the economic and political purposes his portraits and landscapes served in this feudal frontier.

James Wiley Magoffin, a Santa Fe-Chihuahua trader, was the most powerful American in El Paso. He originally moved from Kentucky to Missouri, then ventured west in 1849. His compound, built around a large open square, consisted of several large adobe buildings and warehouses filled with retail merchandise. Magoffin also had a large and impressive mansion of hacienda proportions where he lavishly entertained Bartlett and other government officers, serving them imported French champagne. Part of his house served as Bartlett's headquarters whenever he was in the region. Magoffin named his compound on the north bank of the Rio Grande "Magoffinsville;" it was frequently referred to as "the American El Paso." Important aspects of Magoffin's trade business were provisioning emigrant parties bound for California, and providing supplies and lodging for army troops stationed in the area. Magoffin also promoted real estate in the region. His brother Samuel, another Santa Fe-Chihuahua trader, wrote in a letter: "If a Railroad is made to the Pacific that will pass somewhere convenient to El Paso that property will be worth millions."[39] Their expectations were apparent. Naturally, Magoffin was delighted to accommodate Bartlett's well-funded government party—for a good price.

Pratt's portrait of Magoffin presents the shrewd trader as a statesman of wealth and cultured refinement (Pl. 5). Magoffin is posed by Pratt against a half-open red curtain, through which the viewer sees an open courtyard with a fountain and a hint of blue sky. Magoffin's head is framed against a large column with a smaller column and arches opposite. Typically, adobe hacienda-style buildings of this period did not have columns. Magoffin is represented as a prosperous businessman, wearing a black cravat and a starched white collar shirt, waistcoat, and jacket. His head looks slightly exaggerated with its prominent forehead, alert blue eyes, and intense penetrating gaze. Magoffin's pompadour hairstyle suggests affinities to another famous Southern statesman John C. Calhoun. On 8 October 1852, Magoffin invited Bartlett and "a number of gentlemen to partake of a cold collation, which proved to be one that would have done credit to the caterer of a metropolitan hotel ... the delicacies prepared in New York and Paris for foreign markets can always be found herein abundance, though at a high cost."[40] Doubtless, the visit of a large government party such as Bartlett's was very lucrative for the trader, and he understandably wished to secure Pratt's portrait of the Boundary Commissioner, which he prominently displayed in his house. When Bartlett left El Paso for the last time in 1852, he praised Magoffin's service to the government, and thanked him for accepting his "drafts upon the government" when to have refused them would have caused the Commissioner "serious embarrassment."

Benjamin Franklin Coons, another El Paso trader who profited by supplying Bartlett, was painted by Pratt in the summer of 1851 (Pl. 6). Shrewed and enterprising, the college educated Coons came to El Paso from St. Louis where he had made quick profits in the booming Santa Fe-Chihuahua trade from 1846 to 1848. In 1848 he established a store on property purchased from Señor Juan Maria Ponce de Leon, dubbed Coons Rancho, just west of Magoffinsville.[41] Coons profited enormously by renting his property to the U. S. Army for an outrageous price; it soon became known as "the Post at El Paso," or as "Franklin" from Coons's middle name. For a number of years the American settlement at El Paso was called Franklin in honor of Coons's presence there. When Pratt painted Coons, the entrepreneur was only twenty-five years old and had already made and lost a fortune in the transportation of supplies from the coast of Texas to El Paso. Pratt's portrait shows the youthful Coons in a confident, Napoleonic gesture, sporting a dashing mustache and goatee beard. Behind him is a topographically accurate view of the famed Paso del Norte—route to the riches of California. The portrait signifies both Coons success as a trader and his ambition to forge a transcontinental trading route between the Gulf Coast, San Antonio, El Paso, and on to San Diego in California.

At the same time that Pratt was painting Coons's portrait he was also commissioned to paint his property, *Coons Rancho, Paseo del Norte, Texas* (Pl. 7). In his letter to Bartlett of 23 July 1851 from El Paso, Pratt informed the Commissioner that he had made, "a landscape 'View of this Valley' for Mr. Coons." In a postscript to his letter, Pratt provided important information, "I shall preserve the outline of the view I am painting, here, so as to be able to execute the same view for you if you think it proper. I am painting one now from a point about 1/2 mile above the station, looking down the river, which makes a beautiful landscape."[42] Pratt chose not to paint the town of El Paso that was situated on the Mexican side of the Rio Grande, a much larger settlement of more than five thousand inhabitants, but rather the American side of the river where a small group of Anglo settlers had recently established themselves. Bartlett's *Personal Narrative* clarified the importance of the view Pratt painted as a representation of Mr. Coons's ranch:

> *On the American side there are but few houses; and these may be divided into three groups or settlements. The first is Coons' Rancho. This was the first settlement, and was the military post for about three years.... Many of the buildings are now unoccupied.*
>
> *About one and a half miles below is the principal village, which was established by James W. Magoffin, Esq., a gentleman from Missouri, and one of the oldest American settlers in the country. This place is called Magoffinsville.... This is an admirable situation for a town, and will, no doubt, be the center of the American settlements at El Paso.*[43]

Pratt relied on conventional pictorial constructions in representing the American settlements. In composing his painting, Pratt selected an elevated view of the valley. In the foreground Pratt placed himself sketching. He is gazed upon by a woman wrapped in a blue mantilla and a man who is seated nearby admiring the artist at his work. Behind him are exotic desert plants like the barrel cactus and the ocotillo. To the right of the figures is a classicized tomb or mausoleum of a prominent person. The middle distance in Pratt's image represents the desert transformed into a settled, almost Arcadian landscape with the cottonwood tree-lined *acequia madre* and plowed fields. In the distance is the Rio Grande, and on its opposite bank the numerous haciendas of the Mexican settlement. Beyond Coons's compound, minute figures, perhaps of American soldiers, are moving in formation. They symbolize the political power that enforced American control over the region and held at bay the Apache raids on the valley. They were also the source of Coons's income from rental of the property to the Army. In the compound are five large covered wagons of the type Coons had used to make his fortune, as well as several military style tents such as those used by the Army during this period.

Adapting formulas he had learned from Thomas Cole, Pratt provided the viewer of his painting, in this case the presumptive owner Coons, with an elevated representation of the rancho. To do this he employed the prospect, where by convention the viewer gazes down at a wide expanse of terrain to visually "command" a view of the landscape, in this case stretching out down to the Rio Grande Valley. In the distant sky brilliant sunrise light streams down on the scene—the morning of the new day of American reclamation of this rich valley. In a very tangible sense Pratt's painting provided a visual confirmation for Coons, and for other prominent members of the American community, of their hold on the land, of the value of their recently acquired real estate, and of their place in the unfolding drama of national history.

Yet, an unseen flaw existed in this deceptively simple visual representation of real estate. At exactly the moment Coons was having his portrait and the view of his property painted by Pratt, he was facing financial ruin. In mid-August 1850, Coons, who had already suffered serious reverses when a large wagon train of supplies he purchased to sell to the Army in El Paso was lost, sold most of his remaining wagons, mules, and horses to the notorious adventurer and confidence man Parker H. French.[44] The drafts he received from French on a New York bank turned out to be forgeries. By the summer of 1851 Coons was a desperate man, willing to spend money he did not have to commission Pratt to paint property he was about to lose, and a portrait that belied his impending bankruptcy. Indeed, in July 1851, at the same moment Pratt was painting his property, the Army withdrew from its rental agreement to lodge soldiers at Coons Ranch. By September they were gone and his property foreclosed. It was for this reason that Bartlett wrote that "Many of the buildings are now unoccupied."

The Inez Gonzales Incident

Although the desert landscapes of the Southwest entranced Bartlett, some of his attitudes about the region's inhabitants, both Native American and Hispanic, would be problematic today. No event from the *Personal Narrative* exemplifies these ethnic issues more than the rescue of Inez Gonzales. It was an experience Bartlett thought "to awaken the finest sympathies of our nature; and by its happy result afforded a full recompense for the trials and hardships attending our sojourn in this inhospitable wilderness."[45] Fifteen-year-old Inez Gonzales was a Mexican girl who had been abducted from her family by Apaches who then sold her to a group of white traders. When Bartlett met her they were taking her to New Mexico to be sold as a slave to the highest bidder. Bartlett invoked his authority under terms of the Treaty of Guadalupe Hidalgo prohibiting trade in stolen horses, mules, or humans, and seized Inez with the intention of returning her to her parents. He wrote in his Narrative, "I therefore deemed it to be my duty—and a pleasant one it certainly was, to extend over her the protection of the laws of the United States, and to see that, until delivered... to her parents, she should be 'treated with the utmost hospitality' that our position would allow."[46] Bartlett's company provided clothing for Inez, "but with all the attentions extended to her, her situation was far from enviable in a camp of over a hundred men, without a single female with whom she could hold any intercourse." Bartlett successfully returned Inez to her family in Santa Cruz, Sonora. However, a few months later she was found at the Presidio at Tubac, "to our infinite astonishment and regret," in what appeared to be similar oppressed circumstances, under the control of a Mexican Army officer.

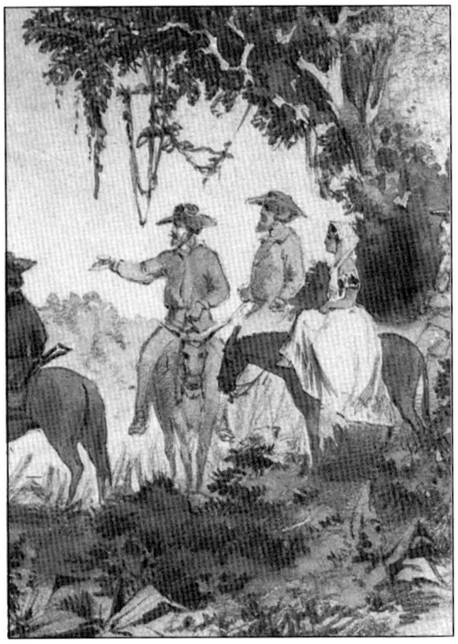

Figure 25
John Russell Bartlett
Inez Gonzales, detail from *Fording a Stream, Packmules Sink in Quicksand*, September 13, 1851
10" x 12 3/4" • Pencil and sepia wash
The John Carter Brown Library, Brown University, Providence, Rhode Island [117]

During the three-month march with Inez, Bartlett depicted her as a proper Victorian lady. In the one drawing that shows her, Inez is mounted sidesaddle on a mule; she wears a mantilla on her head, and a long skirt (Fig. 25). This gesture of good will towards the young Mexican woman seemed to raise much excitement among members of the expedition. When the party reached San Diego, Pratt quickly painted a portrait of Inez Gonzales in oil on canvas. His work was reviewed in the San Diego *Herald*:

> *Mr. H. C. Pratt, the Artist of the Boundary Commission has succeeded in taking the likeness of several Indian chiefs, to be placed in the Indian Gallery of the Patent Office at Washington City. We have seen a beautiful portrait, executed by him, of Inez Gonzales, a handsome Mexican captive girl. It is designed, we believe, to grace the same national collection.*[47]

Evidently the painting was not acquired by the government, because in 1855 it was exhibited at the Rhode Island Art Association, as *Inez Gonzales, A Mexican Girl*, and its owner was listed as J. R. Bartlett.[48] The painting is today unlocated. The entire incident revealed how Bartlett, the proper, respectful Victorian gentleman scholar, constructed himself as a chivalrous champion of a helpless, lower class Mexican woman. The outcome of the incident must have suggested the futility of outsiders attempting to change misunderstood social customs, or impose their own values.

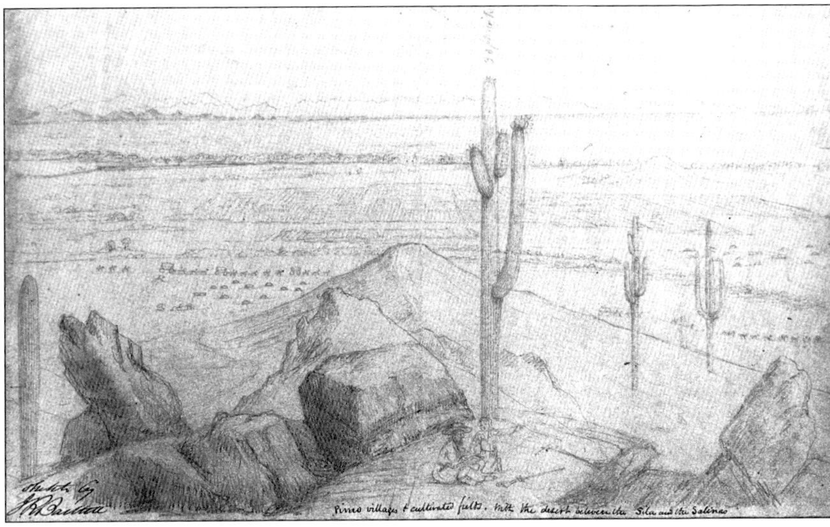

Figure 26
John Russell Bartlett
Pimo [sic] *Villages & Cultivated Fields, with the Desert between the Gila and the Salt* [Rivers], July 3-5, 1852
8 1/2 x 14 • Pencil
The John Carter Brown Library, Brown University, Providence, Rhode Island [72]

NATIVE AMERICANS AND THE RUINS OF THE GREAT HOUSES

Bartlett was an ethnologist fascinated by aboriginal languages, but his *Personal Narrative* is only superficial and anecdotal in its discussion of the more than forty Native American tribes that he encountered. Like most Americans of his period he exhibited a strong preference for tribes that were peaceful and settled agriculturalists, rather than for the nomadic, warlike Apaches. Bartlett believed the Pima-Maricopa people of the Gila and Salt River Valleys were associated with the impressive architectural ruins of Casas Grandes. His enthusiasm for the Pima-Maricopas was abundantly evident in the *Personal Narrative*:

> *There are no tribes of Indians on the continent of North America more deserving of the attention of philanthropists than [the Pima-Maricopa].... None have ever been found further advanced in the arts and habits of civilized life. None exhibit a more peaceful disposition, or a greater simplicity of character; and certainly none excel them in virtue and honesty.... Our Missionary Societies and Boards of Education could not devote a portion of their means to a more worthy object, than to send teachers and mechanics among this interesting people before they become contaminated by intercourse with whites....*[49]

The Gila and Salt River Valleys were particularly important to Bartlett's party because they were thought to offer the best route for a transcontinental railroad, and because the abundant water in these rivers held forth the promise of wide-scale irrigated agriculture. Pratt, who by this point had finally joined the party from El Paso, made a sketch after a drawing by Bartlett of the lands of the Pima-Maricopa Tribes in the fertile Gila and Salt River Valleys, to the west of the location of what is now the city of Phoenix (Fig. 26). In Bartlett's original drawing, and in Pratt's more elaborately worked-out watercolor (Pl. 9), the rich bottom-lands of the valley with their irrigated fields are clearly represented in the distance. The obvious potential for agriculture in this part of the desert was particularly appealing, and the popularity of Bartlett's *Personal Narrative* with settlers soon helped to stimulate settlement. A few decades later, when Anglos began to arrive, one of their first acts was to repair and expand the Native American-built canal system for their own farms. This was a view that Pratt selected to develop into his major oil painting *View from Maricopa Mountain near the Rio Gila* (Pl. 10), which was exhibited in Boston in 1855. Dominating the surrounding desert landscape, and the foreground of Pratt's watercolor and later painting, are towering 30 foot high saguaro cacti. These strange plants suggested to Bartlett the ruins of a majestic palace, the columns alone of which were left standing.

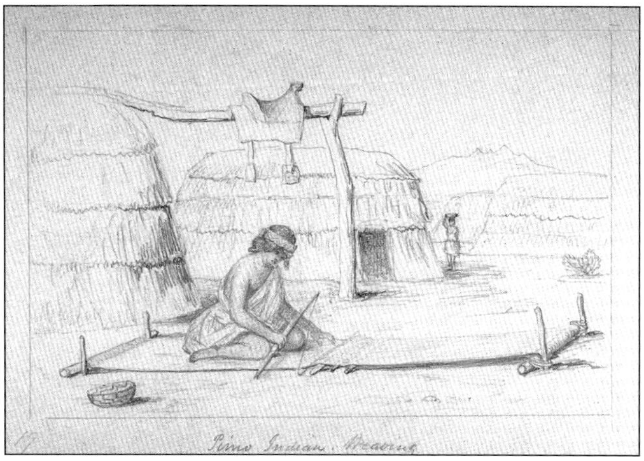

Figure 27
John Russell Bartlett
Pimo [sic] *Indian Weaving*, June-July 1852
4 1/2" x 7 1/4" • Pencil
The John Carter Brown Library, Brown University, Providence, Rhode Island [69]

The "semi-civilized" customs and arts of the Pima Maricopas fascinated Bartlett; he discussed them extensively in his *Personal Narrative*. He drew their weaving techniques, *Pimo* [sic] *Indian Weaving* (Fig. 27), and later commissioned Seth Eastman to produce a watercolor, *Village of the Pimo* [sic] *Indians, Gila* (Pl. 8) from drawings of that are unlocated today. Throughout his sojourn with these tribes Bartlett stressed, both in his drawings and in his text, their exoticism, their adaptability to Anglo ways, and above all, the desirability of their lands. Thus he conformed to the dominant ideology of his period: the Native Americans must either adapt to European-American ways, or be destroyed.

For Bartlett the ethnologist and historian, the most fascinating antiquities of the region were the impressive ruins of Casas Grandes or "great palaces." Legend claimed they were built by Aztecs centuries earlier, and they were sometimes even called Montezuma's houses. Bartlett was eager to dispute this. He and Pratt visited two important sites, one along the Gila River in what is now Arizona, and one farther south in Chihuahua, both of which Bartlett sketched. At the Casas Grandes near the Gila (Pl. 11), Bartlett's company was oppressed by the intense desert heat. The thermometer remained between 110 and 119 degrees, making appreciation of the ruins difficult. Because of his extensive background in Spanish Colonial literature, Bartlett's visit to these much-discussed ruins was a high point of the expedition. After citing the early Spanish explorers such as Father Pedro Font and Father Kino, as well as the opinions of later writers such as Von Humboldt, Bartlett was at pains to challenge the popular notion, perpetuated by his nemesis Emory, that these ruins were of Aztec origin. Relying on a formal analysis of style, he observed that "in [none] of the relics, ornaments, or works of art, do we observe a resemblance between them."[50] However, Bartlett inscribed another set of associations in his drawings. To him, they strongly suggested ruined medieval castles, and he drew them in a manner that even Cole, or his friend Stephens, might have approved. In his representations Bartlett carefully placed figures of the exploring party in the foreground, admiring, excavating, and measuring the adobe ruins (Fig. 28). Bartlett observed of the larger complex in Chihuahua:

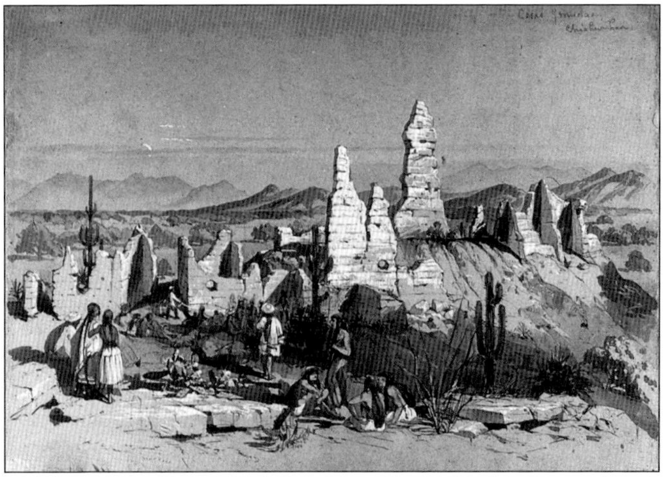

Figure 28
John Russell Bartlett
Casas Grandes, Chihuahua,
August 9-10, 1852
11" x 15" • Pencil and sepia wash
The John Carter Brown Library, Brown University, Providence, Rhode Island [80]

> *I had long known of the existence of these ruins, which are spoken of by various writers on Mexico, and had made frequent inquiries about them, during the winter I spent in El Paso... without getting any satisfactory account of them. All said there were some old adobe buildings there in ruins; but whether they belonged to the present or to an earlier race of Indians, they knew not. At any rate, all agreed that they were not worth visiting.... It required but a glance at these buildings to show what they were, the first being a church, with its ... tower. Remains of tombs were also visible; but from the decayed and weather-worn appearance of the walls, the church had long been abandoned.... I seated myself on a small knoll, or mound, and occupied myself while there remained any light in making a sketch. The Doctor* [Webb] *was engaged in searching among them for relics.*[51]

The appreciation of ruins that emerged most strongly in 18th century European aesthetics was a conventional device for signifying associations of history. More recently, and a precedent that Bartlett and Pratt would have both been familiar with, Thomas Cole had

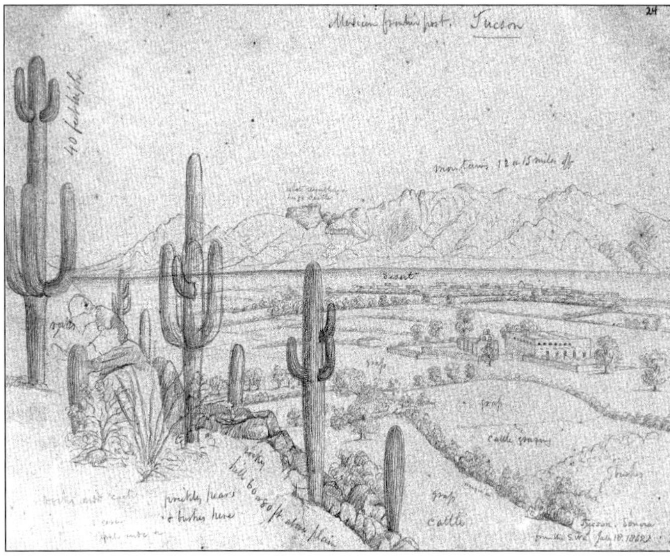

Figure 29
John Russell Bartlett
Sketchbook - Tucson, Sonora from the SW, July 18, 1852
9" x 12" • Pencil
The John Carter Brown Library, Brown University, Providence, Rhode Island [172w]

painted a number of images of ruined medieval castles. Bartlett's invocation of earlier Spanish Colonial writers was yet another means of overlaying associations of the past—of relics, tombs, and abandoned churches—and resulted in a deepening national interest in these mysterious edifices. Unlike his mentor Cole, Pratt concentrated on drawing ruined Spanish Colonial churches. He also elaborated on Bartlett's drawing in his watercolor *Ruins at Casas Grandes, Chihuahua, Mexico* (Pl. 12).

Old Tucson and Spanish Colonial Architecture in the Desert

One of the most striking Spanish Colonial settlements Bartlett's company visited on their way south from the Casas Grandes on the Gila River was the old Presidio of Tucson and its nearby mission churches of San Xavier del Bac and Tumacacori. Bartlett produced impressive panoramic drawings that showed Old Tucson and the distant mountains (Figs. 29 & 30). In his *Personal Narrative* he described the scene precisely:

> *I ascended a rocky hill ... and had a fine view of the valley and surrounding country. This hill was covered with cacti of every variety, among which the petahaya* [Saguaro] *was most prominent. The fouquiera, yucca, and agave, were also luxuriating among the rocks; for scarcely a particle of soil could be seen. Below was the valley, its fields and acequias defined by the long lines of shrubbery and trees. The town is seen on the border of the plateau, where the desert begins, and stretches off to the east, to a high range of mountains about fifteen miles distant. In these mountains is a singular cliff of rock resembling a huge castle.*[52]

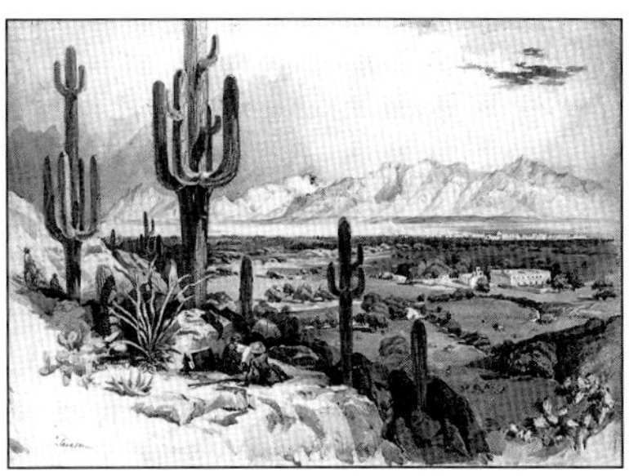

Figure 30
John Russell Bartlett
Tucson, Arizona, July 17-18, 1852
12 3/4" x 18 1/8" • Pencil and sepia wash
The John Carter Brown Library, Brown University, Providence, Rhode Island [75]

In the foreground of one of the drawings, amid the exotic desert vegetation, an artist is depicted sketching the distant vista complete with its dramatic play of light and shadow over the mountains. The scene was permeated with sentiments of history, and the accompanying text implied the necessity of American interventions to control Apache raids on the town. According to Bartlett:

> *Tucson is the most northern town in Mexico, and a very old*

place. It is found on the oldest maps, and is referred to by the early missionaries. It has always been, and is to this day, a presidio or garrison; but for which the place could not be sustained. In its best days it boasted a population of a thousand souls, now diminished to about one third that number.... The lands near Tucson are very rich, and were once extensively cultivated; but the encroachments of the Apaches compelled the people to abandon their ranchos and seek safety within the town. The miserable population, confined to such narrow limits, barely gains a subsistence, and could not exist a year but for the protection from the troops. More than once the town has been invested by ... Indians, and attempts made to take it, but thus far without success....[53]

Bartlett's and Pratt's frequent practice of representing a scene from an elevated point of view to embrace an expansive landscape has recently come under the scrutiny of art historians. Alan Wallach and Albert Boime have shown how ideological constructions of dominion and appropriation were often signified in such panoramic views.[54] Bartlett's frequent recourse to this convention, as seen in his dra-matic view of Tucson, underscores the pervasiveness of this iconological construction. In Bartlett's image, the viewer is invited to share the gaze of the artist-explorer over historic old Tucson and its fertile valley. Such images functioned, as did much of the text of Bartlett's *Personal Narrative*, to support the ideology of American expansionism. As Bartlett observed, "... the early discoverers saw the great advantages of [such] places; although it was left for the Americans one hundred and fifty years later to avail themselves of them."[55]

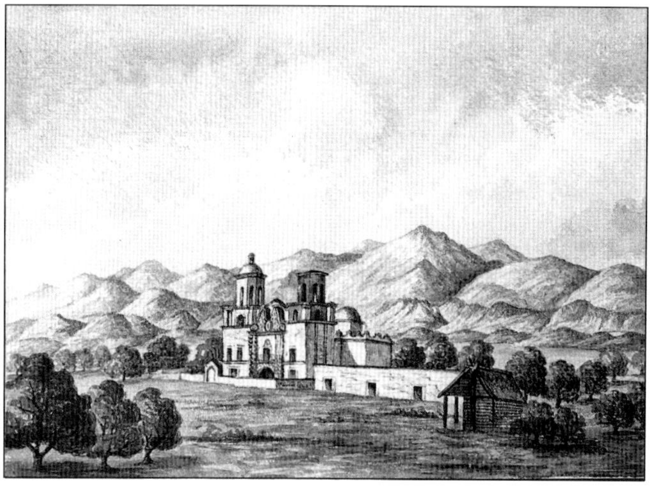

Figure 31
Henry Cheever Pratt
San Xavier del Bac, July 19, 1852
10" x 15" • Watercolor
The John Carter Brown Library, Brown University, Providence, Rhode Island [30]

Spanish Colonial history captivated Bartlett; he was fascinated by the mission churches built across the Southwest. He viewed the failed Spanish empire as a precursor to the era of American enterprise and transcontinental dominion. The most impressive of the missions visited was the Church of San Xavier del Bac, which Bartlett greatly admired, despite its setting (Fig. 31):

A ride of nine miles brought us to the mission of San Xavier del Bac; truly a miserable place, consisting of from eighty to one hundred huts, or wigwams, made of mud or straw, the sole occupants of which are Pimo [sic] *Indians, though generally called Papagos. In the midst of these hovels stands the largest and most beautiful church in the State of Sonora. It is built of brick on the summit of a low hill, and has two towers and a dome. In a square, around and directly connected with the church, are some adobe houses, which were occupied when the Mission was in a flourishing state. All save one are now tenantless....*[56]

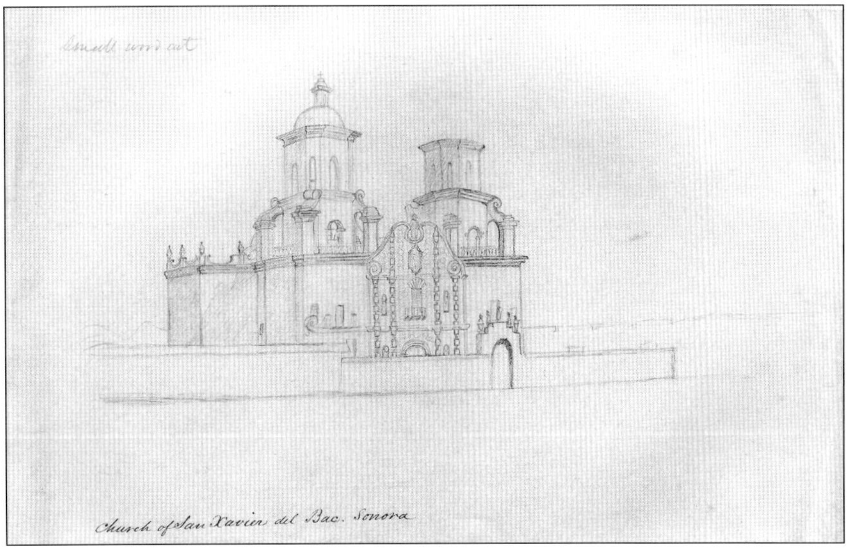

Figure 32
Henry Cheever Pratt
Church of San Xavier del Bac, Sonora, July 19, 1852
5 1/2" x 9 1/4" • Pencil
The John Carter Brown Library, Brown University, Providence, Rhode Island [19]

The elaborate architecture of San Xavier particularly impressed Bartlett, although he recognized the similarity of its architectural plan to many other Spanish mission churches. From his close reading of Alegre's *History of the Society of the Jesuits in New Spain* Bartlett knew that the building dated from the late 18th century, although Father Kino had built an earlier church there around 1700. Its elaborate ornamentation, derived by the Native American artists from study of architectural pattern books, struck him as "new in architecture, partaking neither of the Greek, Roman, nor Gothic orders." Originality was evident:

Along the eaves is a row of queer looking creatures, the like of which cannot probably be found, even in this country of strange animals. The interior is gaudily painted; and from the profusion of gilding, one might suppose the Mission to have possessed a gold placer. Around the altar, and in the niches, are many wooden statues, from the size of a foot to that of life. The poor Indian doubtless believed them all to be saints, and made his offerings accordingly, although about one half are statues of old Spanish cavaliers and figures of Chinese mandarins. There are besides angels and archangels, or figures intended to represent celestial beings with wings, five or six feet high, springing from the walls.[57]

Upon his return to Boston, Pratt produced a painting of San Xavier that has recently been rediscovered (Pl. 13). Although unsigned, it has a strong formal resemblance to other works by Pratt, particularly his watercolor, and pencil sketch of the church facade (Fig. 32). It is also similar in its handling of paint to other works by the artist. Pratt's image conveys something of the desolation Bartlett experienced at the sight of the abandoned church. Its elaborate architectural presence fills up the painting's space almost as if to deny the terrible barrenness of the place.

Farther south, the party encountered the impressive ruins of Tumacacori, that Bartlett pronounced, "beautiful and picturesque." Pratt sketched the church that had formerly been an important mission, and he later produced one of his most impressive oils of the ruins (Pl. 14). The effect of the impressive church, and its surrounding mission compound, in ruins, was made more acute by the party's awareness that the desolation resulted from recent Apache raids. The arrival of American authority in the region promised to end these depredations. Surrounded by a large grove of trees, the mission was situated about a mile from the road. Bartlett did not see the path leading off the main road to the site, and as a consequence missed visiting the place. Pratt, however, saw his opportunity and sketched the church in drawings which he must have retained, because there are no sketches of Tumacacori among the works at The John Carter Brown Library.

In Pratt's painting, which he carefully signed and titled *The Deserted Church of Tumacacori, Estado de Sonora, Mexico*, but did not date, the ruin is framed by its surrounding adobe outbuildings and set amidst a background of trees and distant mountain peaks. The elaborate baroque facade that Pratt carefully delineated derived from Spanish pattern books like San Xavier's did, and possessed the architectural conceit of a broken pediment, dome, and large ruined tower. In the foreground of the painting a few members of the survey party ride by, they include a figure on a white horse. This may be Commissioner Bartlett, who did not see the mission. The position of this figure may also be Pratt's sly comment on the gentleman's rivalry that seems to have emerged between the two artists towards the end of the expedition. By this time, Pratt may have become annoyed at Bartlett who seems to have frequently wanted to be the first to sketch a site. In the distance, the influence of Cole is evident in Pratt's generic, pyramidal mountain forms. The view point is from above the humble thatched *ramadas* where the few Mexican inhabitants of the place lived. Their poverty and the desolation of the once beautiful church, contrasted by the fertility of the surrounding valley, seem to be an argument in favor of the arrival of the Americans to set things right.

The experience of San Xavier, Tumacacori, and other ruined mission churches prompted Bartlett to assert in his *Personal Narrative* the superiority of Protestant English methods of settlement as opposed to those of the Catholic Spanish:

> *What a marked difference there is in Spanish and English colonization! Here the zealous Missionary preceded all others, planting the cross along with the banner of his country. Then commenced the work of baptizing; and as soon as a sufficient number of converts had been made, a fertile valley was chosen, and a church erected with buildings to accommodate some hundreds. Next came the colonists, whose main efforts were to support the Mission and its priests. The Anglo-Saxon pioneer entered the wilderness with his axe, his plough, and his rifle; and after he had erected his own dwelling, the mill, and blacksmith's shop rose up. Lands were brought into cultivation, the mechanic arts flourished; and when the colony became large enough and rich enough to support a pastor, a church was built.*[58]

The result of Anglo American freedom from subservience to religion, Bartlett declared in the spirit of the Protestant capitalist, was that American settlers attained more wealth in a few decades than the Spanish had in two centuries.

A Future Desert Arcadia

The ruins of Spanish Colonial culture, and the dilapidated condition of Spanish settlements, were part of a larger conceptual picture of the borderlands that Bartlett sketched. Often Bartlett's views of a region in decline were in conflict with his desire to promote the economic potential of the Southwest. He believed the solution was American development to sustain the productivity of the land. Bartlett's idea of the future included a landscape that was a potential Arcadia requiring only the civilizing progress of Anglo American commerce and agriculture to achieve its great promise.

Examples can be seen in the foreground of Bartlett's pictures of Tucson (Figs. 30 & 31), where he noted "a large hacienda, which in former times must have been very rich ... is now in a decayed state, and but a small portion of its fine lands are cultivated."[59] When the party traveled to La Magdalena in Sonora, they passed San Lazaro, which Pratt sketched (Pl. 15). Bartlett described it as "a large ruined hacienda, on the banks of the Santa Cruz River, where an extensive soap manufactory was formerly carried on. The buildings were beautifully situated in the valley ... with an extensive orchard of peach and quince trees...." Bartlett walked through the deserted buildings that "seemed to have been built with a view to comfort and convenience unusual in the country. It had been deserted for six years.... Many deserted and ruined ranchos were seen in the valley."[60] A few miles farther and the expedition encountered the ruined church of Cocospera that had been abandoned because of attacks by Apaches. "Proceeding down the valley, we noticed many ruined ranchos, corrals, and other remains of a civilized community, now over grown with tall grass and shrubbery.... It was really sad to see so beautiful a region reverting to the condition of a wilderness in consequence of the attacks of ruthless marauders."[61] As Bartlett surveyed his maps of the region he also experienced sentiments of a great past history, and a mood of somber desolation:

> *Along the valley of the Rio Grande and in New Mexico are a host of names on the map applied to ranchos, and sometimes to places where neither villages nor ranchos now exist.... A stranger on looking at one of these maps would imagine the country thickly settled, whereas there might not be a village, rancho, or even a single inhabitant, where he is treated to a long list of names, including half the saints in the calendar, all the apostles, and the Holy Lady of Guadalupe into the bargain.*[62]

On the other hand, Bartlett held no illusions that the arrival of settlers from the United States was fraught with severe problems for the indigenous peoples. He observed:

> *The quiet people here have been so much annoyed by the conduct of California emigrants who have passed through the country* [Mexico, La Magdalena], *as to make them shy of all Americans. These reckless adventurers often set at defiance all law and propriety, and we had many accounts of their shameful and brutal conduct.... They will enter a house, pistol in hand, demanding whatever it affords;... and commit other outrages which make one who has any national pride blush to hear recited.*[63]

Bartlett harbored an attitude of typical American superiority toward Native Americans and the Hispanic-Americans of the bor-

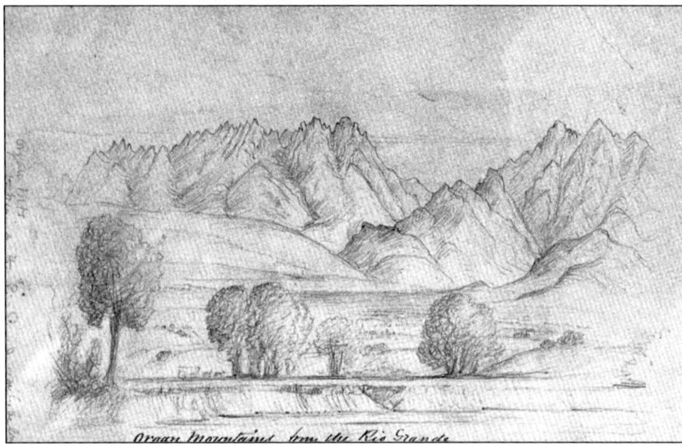

Figure 33
Henry Cheever Pratt
Organ Mountains from the Rio Grande, September 1852
5 1/4" x 8 1/4" • Pencil
The John Carter Brown Library, Brown University, Providence, Rhode Island [24]

derlands, but at the same time was able to condemn the greed and excesses of his fellow Americans. He understood that the arrival of American military, political, and economic power in the region would soon bring irreversible changes, and that some of these would be for the better while others would not.

Desert Landscapes of Sublimity and Wonder

Among Bartlett's and Pratt's many Southwest experiences, few compared to their response to the extraordinary landscapes encountered. Bartlett's narrative is full of references to the vastness, sublimity, and desolation of the region, but it was the fantastic forms of the *Sierra de Los Organos*, or The Organ Mountains, of southern New Mexico, about fifty miles north of El Paso, that elicited his strongest reactions. Both Bartlett and Pratt sketched these mountains that were named from their many jagged pinnacles, which were likened to the pipes of an organ (Figs. 33, 34, & Pl. 16). Bartlett thought the name appropriate because, "both the Spaniards and the aborigines display a much better taste in the appellations given by them to mountains, and other objects of natural scenery, than is usually exhibited by our people. Their names," Bartlett declared, "are significant of the appearance which the mountain assumes, while ours are christened after some military officer or politician, who may have made a little noise in his day, but may have never been near the locality which bears his name."[64] Bartlett was surely thinking of his antagonist William H. Emory who had named a mountain for himself.

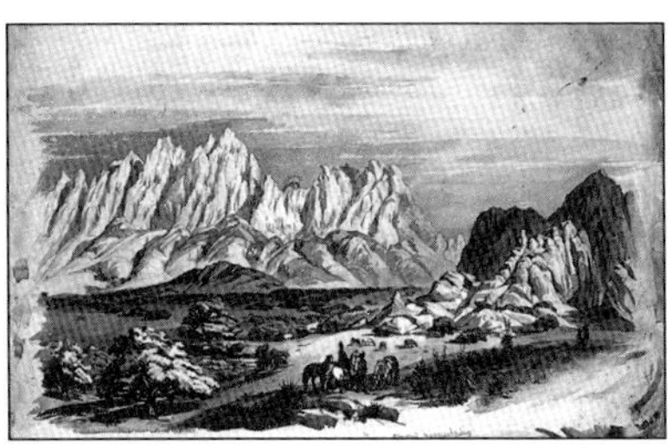

Figure 34
John Russell Bartlett
Organ Mountains, New Mexico, September 1852
13" x 18 1/4" • Pencil and sepia wash
The John Carter Brown Library, Brown University, Providence, Rhode Island [82]

The Organ Mountains possessed historic associations and potential natural resources that made them particularly interesting to the survey party. Because of their great elevation their peaks were covered in heavy pine forests, while their lower valleys held groves of oak. Since the region near El Paso was largely without good timber for building, Bartlett thought these stands of trees would be valuable to future settlers and railroad builders. There was also promised mineral wealth. When Bartlett returned to El Paso from California in 1852 he made a point of revisiting the range to inspect a recently opened silver mine. The foreboding mountain range was known as a haunt of the Apaches so the American garrison at El Paso had recently been moved to Fort Fillmore, a new outpost in Southern New Mexico. Bartlett and Pratt were able to enter the mountain

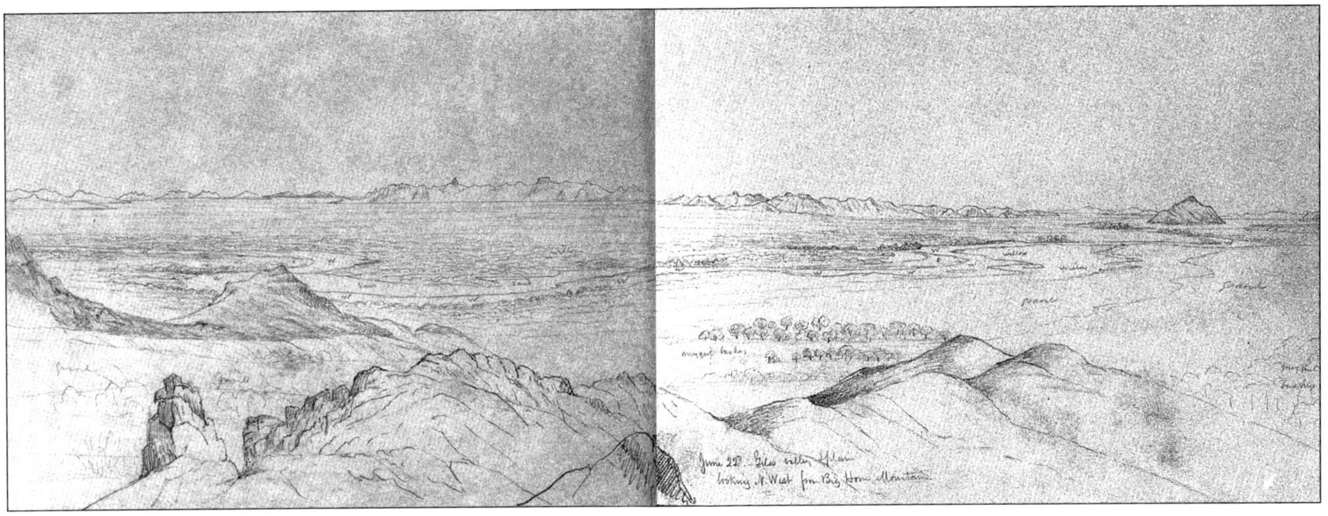

Figure 35
John Russell Bartlett
Sketchbook - Gila, 1852
9" x 12" • Pencil
The John Carter Brown Library, Brown University, Providence, Rhode Island [173]

range only under the protection of an escort of dragoons from the fort. After exploring the mountains the party saw evidence of Indian presence and quickly withdrew. "From the place where we halted and lunched, I took a sketch of these mountains and of the defile through which I had passed," Bartlett wrote. "A small stream flowed near us, marked by a line of fine large oaks. Midway between this spot and the mountains rises a bold mass of white granite."[65]

When he returned to Boston, Pratt produced an oil painting, *Grand View of the "Organ Mountains,"* which was described as having a small camp of Apaches in the foreground.[66] This painting is unlocated today. Given the appearance of both Pratt's and Bartlett's striking sketches it must have been one of the artist's most important images of the magnificent scenery of the Southwest.

Pratt and Bartlett were enthralled by the spectacular Gila River Valley, the proposed route for a transcontinental railroad. Both artists produced panoramic sketches of the scenery. Pratt's 1855 oil painting, *View From Maricopa Mountain,* prominently displayed his reliance on conventions of panoramic representation that he had absorbed from Cole (Pl. 9). Bartlett made a pencil sketch of the *Gila Valley and Plain, Looking Northwest from the Big Horn Mountains* that spread across two pages of his sketch book (Fig. 35). Bartlett's pencil drawing encompassed many square miles of river valley and distant mountain ranges and his text expressed his excited response to the place:

> *The mountains here are as desolate and barren as it is possible to conceive. Not a tree or a shrub could be seen on them, while their bold and abrupt sides are furrowed with huge chasms and gorges. Between the base of the mountains and the bottom-land are low gravelly hills covered with the Spanish bayonet,*

agave, and various kinds of cacti.... The Gila here widens considerably, and is proportionably [sic] *shallow and filled with sand-bars.*[67]

Pratt also relied on panoramic constructions of the landscape, as viewed from a commanding perspective. Yet unlike his mentor Cole, who struggled with deep intellectual conflicts about American progress, Pratt was evidently something of an opportunist, eager to produce art in support of the doctrine of Manifest Destiny. His watercolor *View on the Gila*, with its suggestion of a fertile riparian river landscape (Pl. 17), was at odds with Bartlett's bleak depiction of the same scene. After traveling all night to avoid the intense heat, "We now continued another half mile near the base of a black-looking rocky bluff without a tree to shade us." Describing the view taken in Pratt's watercolor, and in an oil on canvas that Pratt later produced, Bartlett observed, "The bottom-land, or valley, which is visible from the summit of this bluff for twenty miles, is altogether sand, with a few clumps of willows on the margin of the river. Not an acre of arable land is visible."[68]

Pratt transformed this watercolor sketch into an oil painting known today as *Basin of the Rio Gila* for his 1855 exhibition in Boston (Pl. 18). He raised the horizon line in the painting to further accentuate the panorama of the vast landscape, and placed a summer rain shower on the right. By inserting more small figures, animals, and wagons from the expedition in the middle distance Pratt was able to better suggest the inhuman scale of the region. The figures are shown crossing the river and winding their way across its sandy bottom. In the foreground, tents have been pitched under a menacing rocky repoussoir that frames the left side of the picture. Encoded in the profile of this dark rocky form is an emblematic personification of a human visage.

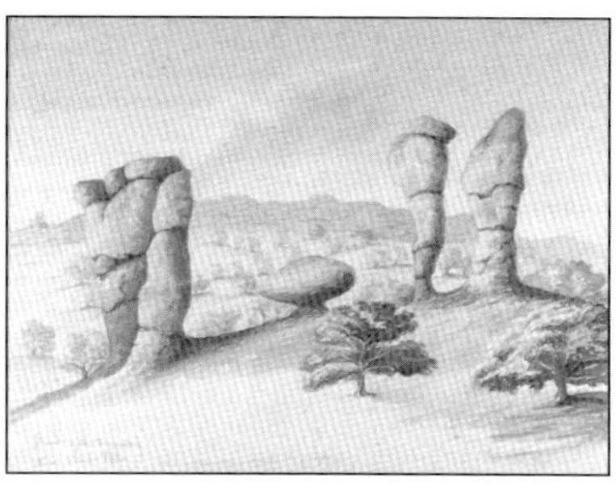

Figure 36
John Russell Bartlett
Giant of the Mimbres near Santa Rita, May 1, 1851
9 1/2" x 13" • Sepia wash
The John Carter Brown Library, Brown University, Providence, Rhode Island [37]

Humanizing the Desert Landscape

Bartlett and Pratt enthusiastically indulged in personifications as they attempted to construct associations that would enable their eastern audiences to comprehend the wild and foreboding south-western landscapes. The anthropomorphic fiction of using objects in the landscape as symbols of human presence was a well-established convention among artist-explorers in the West. Along the Mimbres River in New Mexico Bartlett found:

some singular masses of sandstone standing detached from the adjacent hills, one of them bearing a curious resemblance to a man. My timid mule was much alarmed at the gigantic object.... Around us stood these singular isolated rocks, some appearing like castles, others like single pedestals and columns. The one resembling a human figure ... I christened the "Giant of the Mimbres."[69]

He titled his sketch *Giant of the Mimbres* (Fig. 36). Today these unusual geological formations are protected as a New Mexico State Monument called The City of the Rocks. Bartlett frequently noted a variation on the idea that rocks or distant mountains could suggest associations with castles or other grand architectural forms. Near the Waco Mountains east of El Paso Bartlett sketched some piles of rocks that appeared like the ruined walls of a great edifice (Fig. 37). The use of anthropomorphic, or emblematic personifica-tions of natural history were important parts of the vocabulary of symbolism that Bartlett and Pratt relied on to make the strange landscapes they encountered comprehensible.

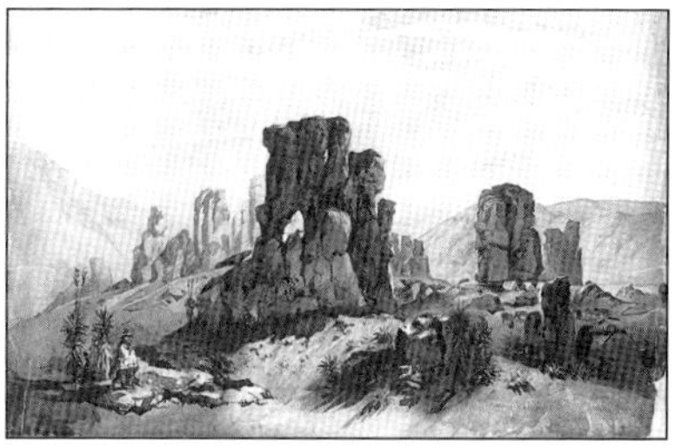

Figure 37
John Russell Bartlett
Granite Masses, Waco Mountains, Texas, March 29, 1851
13" x 18 1/4" • Pencil
The John Carter Brown Library, Brown University, Providence, Rhode Island [102]

OF MINES AND RAILROAD ROUTES: EASTWARD THE WEALTH OF EMPIRE

The two major concerns that ran throughout Bartlett's narrative were the discovery of mineral wealth in the regions they explored, and the need to ascertain a route for a transcontinental railroad. Bartlett observed enthusiastically in the introduction to his *Personal Narrative*, "The time is not far distant, either, when crowds as large as those now pressing on to California and Australia will be 'prospecting' among the mountains of Texas, New Mexico, Chihuahua, and Sonora, attracted by similarly rich mineral deposits, and probably with the like splendid success."[70] The mines at Santa Rita del Cobre in New Mexico held the most potential for mineral extraction. This place had long been known to the Mexicans who worked its extraordinarily rich copper and gold deposits from 1804 to 1838, before the mines were forcibly closed by Apache raids.

Bartlett's anxious state of mind as he left the relative security of El Paso for the deserted mines in New Mexico was vividly apparent:

A wild and barren region lay before us, destitute of inhabitants save hostile bands of Indians which roamed over the deserts or hid themselves in the fastnesses of the mountains; where water was exceedingly scarce, where there was but little sustenance for our animals, and where we could expect no assistance in case our provisions fell short.[71]

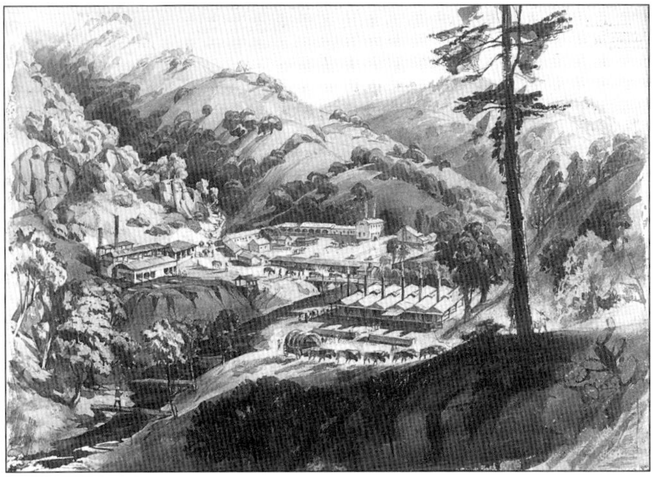

Figure 38
John Russell Bartlett
Quicksilver Furnaces, New Almaden, April 4-5, 1852
11 1/2" x 16" • Pencil and sepia wash
The John Carter Brown Library, Brown University, Providence, Rhode Island [60]

When he arrived at the copper mines, in great pain, suffering with a fractured arm after being kicked by a mule, Bartlett was startled by the desolation of the place; the buildings were in ruins and the mine shafts were caved in. Although no Bartlett drawings of the mines are located, Seth Eastman's watercolor after Bartlett's unlocated drawings of *Bufa del Cobra* [sic] [Santa Rita] shows the ruined compound and buildings (Pl. 21). Upon his recovery, Bartlett expounded enthusiastically about the great potential of the place. Its mineral wealth of copper, silver, gold, and lead was abundantly evident, but most important, Bartlett thought, was its rich supply of timber. "So rich a timbered country does not exist between the Mississippi valley and the Pacific, except in the mountainous district of Upper California," Bartlett declared. "Should a railway be constructed across the country south of the Gila, its timber must be procured from this quarter."[72] The area had fertile lands that could support the miners, although Bartlett worried that until the railroad was constructed transporting the ore would be impossible. As for labor, Commissioner Bartlett declared: "Labor is cheap and abundant in Mexico. At El Paso, Mexican laborers could be had for 62 1/2 cents a day ... but men could doubtless be procured at even a less price. They require only the most simple food: flour, beans, and a very little meat will satisfy their wants."[73]

The copper mines were so important that Pratt produced an oil painting of them for his 1872 exhibition in Boston. The picture, entitled *View in the Southern Part of New Mexico,* is unlocated today. Pratt's exhibition catalog promoted the picture as a depiction of "A very picturesque region, with a delightful climate. Here are found in abundance Copper ores enriched by Silver, and Gold is mined five miles distant. The Southern Pacific Railroad will pass within a few miles of this valley."[74]

In addition to his responsibilities as boundary survey Commissioner, Bartlett's *Official Instructions from the Department of the Interior* specifically ordered him to collect "information relative to the precious metals, quicksilver, and the various mineral ores, and other substances, useful in the arts."[75] As he did with his other instructions, Bartlett interpreted this liberally, making a special visit to the quicksilver mines at New Almaden, California, far from the borderlands. Bartlett sketched the exterior of the mines and the "picturesque scenery in the vicinity" (Figs. 38, 39). The meaning inscribed in these drawings was proof of the superiority of American enterprise and technological expertise over that of Spain. The existence of the mine was powerful evidence of the great riches to be extracted from the west.

The California mines at New Almaden were named for those in Spain, and Bartlett, the historian, was quick to point out the Spanish mine had been in production since Roman times. Until the discovery of the ore in California, the Spanish mine enjoyed a monopoly on the production of mercury, widely used in the production of vermilion pigment—a color used by artists for the depiction of sunsets. Bartlett related the history of quicksilver mining in his narrative and commented on the ever-increasing production and value of the California mine. The development by American entrepreneurs and engineers of new techniques of refining the ore also interested Bartlett. In his drawing of the mine buildings Bartlett visually inventoried the refining process. His drawing of the ascent to the mines highlighted the engineering feat of excavating a road to the top of the mountain to gain easier access to the mine entrance. Bartlett was, however, defensive in justifying the exposure of men and animals to the toxic substance. "Knowing the effects of mercury on the [nervous] system, the question will naturally arise ... as to whether those employed in the mine or about the furnaces suffer from their close contact with the ore or the quicksilver."[76] He quickly added that although the laborers refining the ore salivated heavily from exposure, only horses and mules were known to have died from mercury poisoning.

Bartlett was especially preoccupied with locating a suitable route for a railroad to transport this new found mineral wealth of the West to eastern markets. His *Official Instructions* were specific on this point: "By the sixth article of the Treaty of Guadalupe Hidalgo, provision is made for the collection of information relative to the construction of a 'road, canal, or railway, which shall, in whole or in part, run upon the river Gila....'"[77] A railroad was essential to opening the region to entrepreneurs, settlers, and tourists. Throughout his narrative Bartlett commented on the suitability of the region for a railroad. In the notes for the book he concluded:

> *It will be seen that the advantages of the southern route for a railway, are an open and remarkably level country from the Mississippi to the Sierra Nevada of California—a summit*

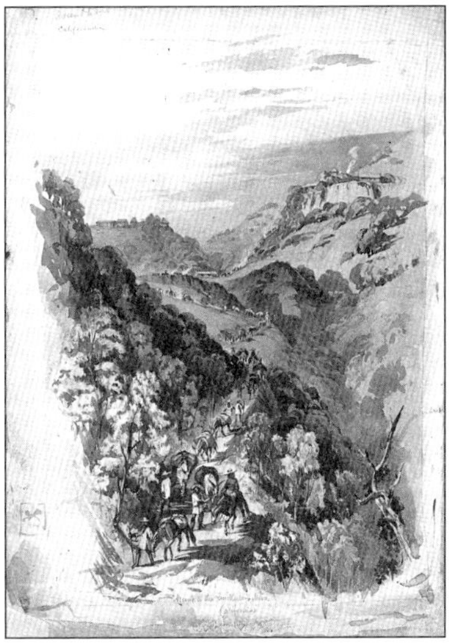

Figure 39
John Russell Bartlett
Ascent to the Quicksilver Mines, California, April 4-5, 1852
17 1/2" x 12 3/4" • Pencil and sepia wash
The John Carter Brown Library, Brown University, Providence, Rhode Island [61]

level a thousand feet less than that of the other routes (so far as known)—entire freedom from snows—and the convenience in obtaining supplies.[78]

When he returned to the East, Bartlett wrote a lengthy commentary on the feasibility of the railroad at the request of the president of the Atlantic and Pacific Railroad, in which he discussed possible routes.[79] Eventually, after the Civil War, the Southern Pacific Railroad was constructed largely according to Bartlett's recommendations, from El Paso along the Gila River and on to San Diego.

The Reception of Bartlett's Book and Pratt's Paintings

After being dismissed as Commissioner of the boundary survey, in 1853, Bartlett set about vigorously defending his work and attempting to publish his findings. He was sandbagged by his political enemies from the survey and by Democrats in Congress. Although he was given strong support by the Texas hero General Sam Houston, who was by then a senator, Congress refused to fund publication of Bartlett's report. Fortunately, with his many contacts in the publishing world, he quickly secured D. Appleton. The *Personal Narrative* was released in less than a year. The illustrations for the volumes had to be greatly reduced in size and scope, and Bartlett did not use many of Pratt's elaborate watercolors. Defensively, Bartlett claimed, "Many of these sketches are panoramic views embracing wide districts of country, and convey to the mind a better idea of it than the most elaborate descriptions." To reduce them to the size of an octavo page, "would detract too much from their merits."[80] Bartlett was also unable to use the superb watercolors he commissioned Seth Eastman to produce, and it was several years before he paid Eastman for them. He continued to cherish "the hope that Congress will yet decide to place the whole results of the Commission before the public in a suitable manner."[81] Perhaps he intended to reserve these images for this final report, which was never realized.

Despite the limitations of its reduced format and lack of color illustrations, the reception of Bartlett's book was overwhelming and positive. The *Boston Post* thought, "The style in which the work is written is lively and animated, and totally devoid of all affectation and straining after effect. The value of the well printed volumes is greatly enhanced by the numerous and well executed engravings with which they are illustrated.... A few revolutions of the wheels of the cars of time," the *Post* asserted, "will almost invariably find parts of, if not all Sonora and Chihuahua an integral part of the United States."[82] Among the Bartlett papers are more than seventeen pages of favorable reviews by newspapers in Boston, New York, and Bartlett's hometown, Providence. The president of the Royal Geographical Society, "pronounced it the best book that had been written upon the subject, and Charles Dickens bears testimony to the interest of the personal narrative."[83]

Buried among these reviews are comments that throw light on the basis for the conflicts between Emory and Bartlett, who represented profoundly conflicting political viewpoints and professional interests. The *Boston Daily Advertiser* noted that the differences between Emory and Bartlett stemmed from "political feeling," and continued that "a great part of the difficulty [lies with] the singular jealousy which military men appear to have of civil superintendence." Following the Gadsden Purchase, Bartlett's old rival Major Emory was appointed to direct a new survey of the border. Emory's final *Report on the United States and Mexican Boundary Survey* was everything in scale and publishing munificence that Bartlett's *Personal Narrative* was not. It appeared in three large quarto volumes with numerous steel engravings and color plates. Newspapers sympathetic to Bartlett's endeavors railed against its exorbitant publishing cost of over $223,000. One critic declared, "It would be nearer to the truth to say the work [Emory's *Report*] cost about three times as much as all the books in the Astor Library!"[84] The same reviewer continued:

> *With the money expended on this work, which it is pretty safe to say would have gone begging for a publisher ... what might have been done for the Library of Congress* [or] ... *the Smithsonian Institution! And yet with all this prodigality, Congress refused to print in the simplest form the really valuable report of Mr. Bartlett.... This is but an illustration of the influence which military men have exerted in our government....*[85]

The bitter controversy between Bartlett and Emory continued on to the late 1850s, by which time Bartlett had been elected Rhode Island's Secretary of State. He was reelected to this office annually until 1872. After the publication of his *Personal Narrative* Bartlett apparently gave up painting. His interest in art and science was replaced by his growing interest in history and bibliography; no documentation of contact with Pratt following the boundary survey exists. Books continued to pour from his pen, although none concerned the Southwest. Bartlett died in Providence in 1886.

Henry Cheever Pratt, in contrast, played out the last acts related to the cultural work of the U.S.-Mexico Boundary Survey. His production of paintings of the Southwest, and their exhibition in Boston in 1855, New York in 1858, and again in Boston in 1872, provided Easterners with images that complemented and augmented Bartlett's popular book. Unlike Bartlett, who never returned, Pratt traveled west again in 1870 following the completion of the transcontinental railroad, attempting to capitalize on the ever-increasing fascination with the West following the opening of the railroad.[86] Although there is no documentation of this later trip, Pratt painted several scenes of the Yosemite Valley, which was not visited by Bartlett's party in 1852.

Pratt's first major exhibition of art related to the boundary survey opened in Boston in late November 1855 at his studio in Armory Hall. The only direct evidence for this is three positive reviews that appeared in the *Boston Evening Transcript,* written by a critic using the pseudonym "Sigma," which praised the work for being "descriptive of the scenery and vegetation, in a portion of our vast domain, with which a very large part of our people are quite unacquainted."[87] Sigma went on to discuss the circumstances of Pratt's engagement with Bartlett's survey: "Under the protection of the government party, he had the most ample opportunity for executing numerous sketches of the novel and highly interesting scenes, which constantly arose in their path." Sigma's reception of these images underscores their importance in making better known a region that few had ever seen:

> *The sketches upon which these paintings are based, are memoranda, faithfully prepared, of the character of the country, its strongly marked peculiarities, and some of its extraordinary productions. How few can have any idea of the vast primitive forests of the Guadaloupe* [sic]*—or the Table Mountains, wearing the appearance of immense fortifications, on the Rio Pecos—or the barren but eminently picturesque and mountainous regions of Texas—or the arid desert ... along the Rio Grande—or that tumultuous range of lofty mountains, upon the Coppermine, in New Mexico.... Few things can be more interesting to such of us as are not permitted to behold the actual scenes, that linger over the graphic representations given, by this accomplished artist, of scenes in Sonora, and Chihuahua; or those which he sketched ... along the Rio Gila, to its junction with the Colorado.*[88]

Sigma concluded with a plea for Pratt's sales: "Go—look at them, and judge for yourselves—and give rapidity to the action of the [auction] hammer, under which they are to be sold—not sacrificed, I hope."

Although this 1855 exhibition was not documented by a catalog, the 5 December review in the *Transcript* makes clear that this showing was the first appearance of Pratt's important painting, *View from Maricopa Mountain, Near the Rio Gila* (Pl. 10). The reviewer for the *Evening Transcript* singled this work out for special praise. "Some of the natural productions, introduced into these paintings," Sigma declared, "are so strange, so colossal, that nothing less than the testimony of an honorable man would entitle the assertion of their real existence to any credit." Dominating the foreground of Pratt's painting is the "great Mexican Cactus." It is an "enormous production ... some fifty feet in height, and from three to four feet in diameter, [as] represented in some of these admirable landscapes, by Mr. Pratt."[89] In conceptualizing his painting Pratt may have been thinking of Bartlett's commentary on the party's difficult trek across the desert to reach the Gila River Valley: "The great cereus [saguaro] here raises its lofty head above all other plants, attaining its greatest perfection in this barren and desolate region. We passed several of gigantic dimensions, and others ... exhibiting singular contortions ... thousands of these cacti rise to the height of twenty or thirty feet, in a single stem without a branch...."[90]

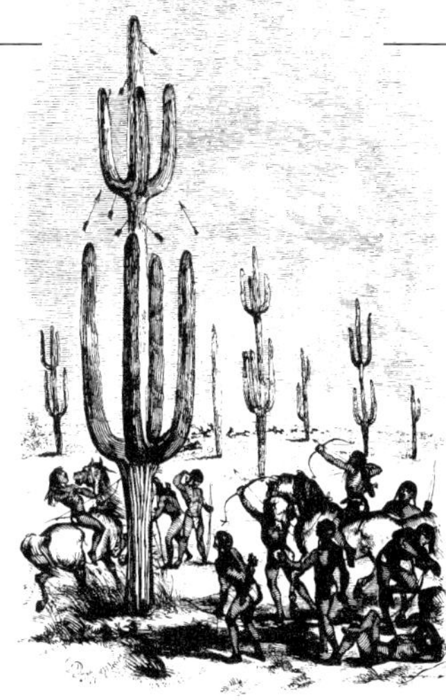

Figure 40
John Russell Bartlett
Indian Amusements, Shooting at the Petahaya
from Bartlett's *Personal Narrative*

From a panoramic perspective the viewer of Pratt's painting gazes over a vast and fertile river valley inviting settlement. In the foreground an Indian arrow dramatically protrudes from the saguaro. Readers of Bartlett's book would have immediately recalled his description of the Pima Indians' favorite amusement of shooting at the huge cactus, and a woodcut illustration (Fig. 40) showing this sport. "In crossing the great plateau, where these plants abound," Bartlett observed, "it is common to see them pierced with arrows."[91] The giant cactus had been a subject of fascination by the survey party, and Pratt made a drawing of the flowers and fruit of the plant that was later adapted for his painting (Fig. 41). In the final painting he represented the full cycle of the cactus's flowering during its most colorful season of growth, late

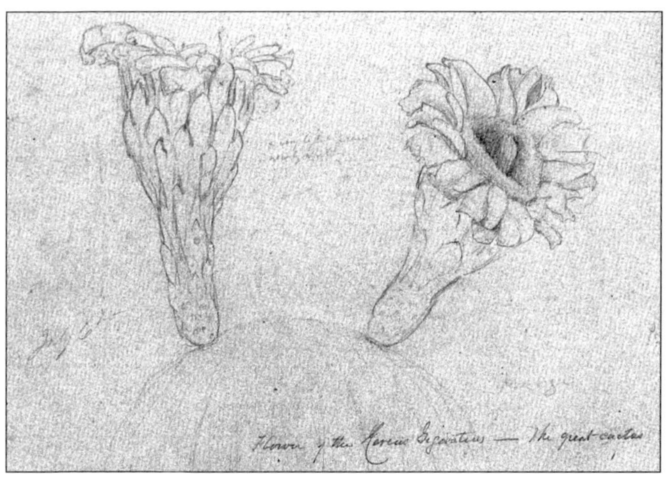

Figure 41
John Russell Bartlett
Flower of Cereus Giganteus - the Great Cactus, June - July 1852
8 1/2 x 13 1/4 • Pencil
The John Carter Brown Library, Brown University, Providence, Rhode Island [149]

spring. The dramatic yellow flowers are shown giving way to seed pods, which in their turn are shown ripening and eventually spilling their red seeds on the ground. Below the plant in the right foreground, another large specimen is depicted with an arrow or spear in its side, while other smaller examples carry the spectator's gaze downward into the distant Gila River Valley. Near the smaller saguaro are two diminutive Native American figures. One appears to have a knapsack to gather the fruits of the cacti, the other carries a staff or arrows similar to those piercing the nearby saguaros. Beyond them, at the foot of the hill, the spectator sees the simple huts of the Pima-Maricopa peoples. In the distance, patches of green suggest the large irrigated fields of the natives which had expanded in size due to the demand for grains from immigrant trains passing westward to California over the preceding decade. The Gila River is depicted full of water with numerous cottonwood trees along its bank. They cut a verdant swath through the otherwise arid landscape. In the left foreground, Pratt inscribes in the profile rock facing the lower saguaro a grimacing, mask-like face, suggesting an ancient aboriginal presence.

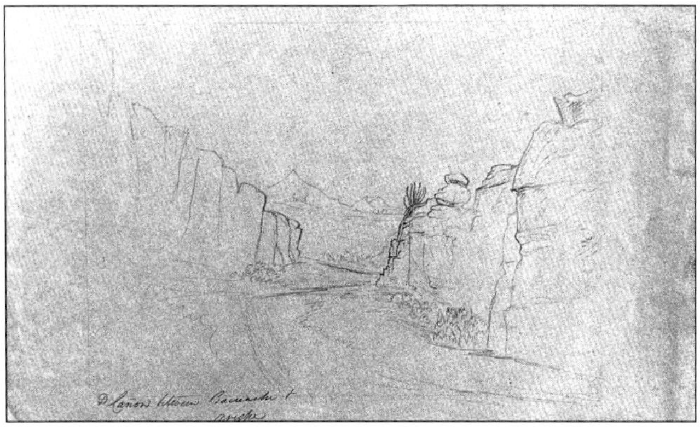

Figure 42
John Russell Bartlett
Cañon between Bacuachi and Arispe, May 30, 1851
8 1/2" x 14" • Pencil
The John Carter Brown Library, Brown University, Providence, Rhode Island [109]

Evidently, this picture was hard to sell because Pratt exhibited it three years later in 1858 at the National Academy of Design in New York under the title *Cactus Cereus, Gigomtus* along with other works, all listed as *"For sale."*[92] It was still for sale in 1872, when the artist, hoping to capitalize on the recent completion of the transcontinental railroad, added a descriptive text to his exhibition catalog. "On the plains below, are seen grainfields and the dwellings of the Maricopa Indians. The new Southern Pacific Railroad will pass by the foot of this mountain." Although the real estate in the picture might be highly desirable, the painting still could not be sold.[93]

Another painting from 1855 is the work known today as *Bartlett Survey Traversing a Canyon* (Pl. 19). Like most of Pratt's images this work builds its meaning through the layering of associations. The painting, however, might better be titled *Bartlett Party Traversing a Canyon Between Bacuachi and Arispe, Sonora.* It is based on a drawing of the same scene attributed to Bartlett and bearing this inscription (Fig. 42). The subject was evidently of special attractiveness because Bartlett commissioned Seth Eastman to prepare a watercolor from this drawing (Pl. 20). The march of the party, self-represented in this image, was filled with great difficulty because of the stifling heat. A desert fire left the landscape "black and gloomy." There was grave danger as well, for only a few miles from the place depicted over a hundred emigrants had been killed by Indians within the preceding two years. Pratt inscribed a sense of foreboding in his image, along with the sense of awe that was felt as they passed through the dramatic wilderness canyon of the Sonora River. The party had to cross and recross the shallow river, always aware that they were under surveillance by Indians. Bartlett's description might be a text Pratt referred to as he produced his picture, which became in his hands an image of a wilderness Garden of Eden:

> *Sometimes for miles we were so closely hemmed in by the perpendicular sides of this extraordinary defile, which rose six or eight hundred feet above on either hand, that we could not see a hundred yards before or behind us.... The whole course of the river through this cañon affords a series of most delightful scenes.... The rocks through which the stream has forced its way, exhibited the most picturesque and fantastic forms. Columns, turrets, towers and pyramids, as nature made them, decked with brilliant flowers or bearing strange cacti, appeared at every turn. From projecting ledges sprung the yucca and agave.... The air was filled with a delicious perfume from the grape and mellilot; and birds of brilliant plumage and sweetest song flitted across our path.*[94]

Pratt's indulgence in anthropomorphic personification is strikingly evident in the geological formation in the right foreground, strongly marked with the suggestion of the classic emblematic profile of a human visage. This face seems to gaze out silently, impassively, at the passing explorers.

Perhaps the most unusual painting in Pratt's 1855 exhibition was a work inscribed on the reverse as *Allegorical Design Representing Evening with a View from Nature Painted on the Gila River, Arizona* (Pl. 22). The painting was based on the experience of a terrible storm the party encountered on the evening of 13 July 1852. Bartlett described the dramatic event that Pratt also witnessed:

> *As the sun sank below the horizon, the dark cloud bank which we had observed far to the south ascended, and we could see the rain already falling on the distant mountain. Night now*

set in; the thick clouds rose higher and higher.... Shrouded as we were in darkness, it was no easy matter to find our way.... But nature's light-house opened its portals, and the vivid lightning flashed around us.... Peals of the most terrific thunder burst upon us.... Next came violent gusts of wind, accompanied by clouds of sand and dust, reminding one of the African simoom. The wind was from the south, and brought the sand directly in our faces.... Lastly came torrents of rain, and this terrific storm was at its height.[95]

The landscape of the image is a conflation of stock elements from several paintings by Cole and from related paintings of the Southwest by Pratt. The inclusion of the flowering yucca plant in the right foreground and the meandering Gila River give an effect of Southwestern authenticity. In the distance, the mountains recall forms found in Pratt's painting of the ruined mission church of Tumacacori. The protuberant shape at the top of the great cliff at the left recalls emblematic profiles from many works by Cole and other of Pratt's contemporaries.

Allegory makes its startling appearance in the desert storm cloud. This swirling vortex of clouds is infused with covert symbolism. Such anthropomorphic symbolism would have been readily comprehensible to Pratt's audiences long familiar with similar symbolism in Cole's painting. In the third painting of *Manhood, from The Voyage of Life,* Cole painted three faces hidden in the clouds representing the vices of suicide, murder, and intemperance. These faces, which Pratt would have known well, provided a precedent for the symbolism in the clouds of Pratt's picture. At the top of Pratt's storm cloud is a figure that looks like a phantasmagorical griffin, while below it, etched faintly in the clouds, are other grotesque faces. Pratt's allegory of evening represents the Southwest as a fantastic landscape filled with mysterious presences where the drama of national destiny unfolds.

Seven years later in 1862, in the middle of the Civil War, Pratt produced another allegorical picture entitled *Evening Scene on the Gila River, Arizona* (Pl. 25). Here the influence of Cole's *Voyage of Life* is unmistakable. The figures of winged angels strewing flowers and floating above a landscape of the Gila River Valley, Pratt's commentary on a nation convulsed in Civil War, recalls Cole's angelic figure that appears in all four panels of his series. The figure of the angel derives from the first two panels of Cole's series, as does the meandering river valley with its luxuriant palm trees and many flowering desert plants, including the giant saguaro. The landscape symbolizes the abundant potential of the West, now deferred by the trauma of war. More pessimistic sentiments are symbolized by the lugubrious reddish light and red clouds. They suggest somber emotions of conflict and death. Pratt's *Evening Scene on the Gila River*, by the time of its execution, is an old-fashioned work, more appropriate to visual strategies and rhetoric of the 1840s. This painting, too, was difficult to sell, because it appears again for sale in Pratt's 1872 exhibition, accompanied by a predictable text, "The Landscape part being a View on the Gila River, on the proposed Route of the new Southern Railroad." Considering the number of times Pratt relied on the proximity of the railroad as an incentive for selling his landscape paintings, the importance of capitalizing on the "unseen hand" of progress becomes evident.

Pratt continued to hope for success with his now out-dated images. In 1862 he produced a pair of paintings entitled *Morning*, or *Aurora*, and *Night* (Pl. 23, 24). His 1872 catalog states that *Morning* is "from the Design in Marble, by Thorwaldsen the Dane." *Night* was described as an "Allegorical Picture, representing Night, after a Design in Marble by Thorwaldsen."[96] Evidence of the prices Pratt received for such pictures is found in an auction notice for "The Artists Sale at the Gallery of Mesrs. William & Everett," where *Morning* was listed as bringing $41.[97] The distant landscape in *Morning* resembles the generic scene in Pratt's painting of the *Allegorical Design Representing Evening...on the Gila River*, but the building in the foreground is not a mission church as has sometimes been claimed, but more likely an Italianate Gothic villa, derived from Pratt's now unlocated painting of *The Valley of San Matteo* [sic], taken from Howards Hill. Pratt's 1872 catalog described the scene of this lost work as "twenty miles (by Railroad) from San Francisco, and contain[ing] some of the most magnificent Villas and Gardens in the country, being Residences of some of the most opulent citizens of San Francisco." *Dawn*, it would appear, represents the West, while Night, with its shooting comets in the sky, and angel carrying sleeping, or possibly dead, children is probably intended to signify the East, plunged into darkness by war. There were two celebrated occurrences of comets in 1861 and 1862, both widely interpreted as signs of Civil War.

Pratt's allegorical figures may have had some influence on the formulation of John Gast's famous 1872 painting and chromolitho-

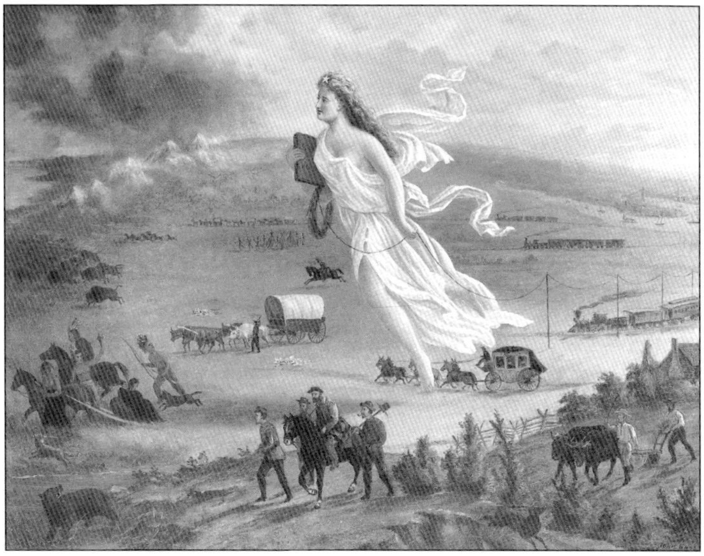

Figure 43
John Gast
American Progress, 1872
12 3/4" x 16 3/4" • Oil on canvas
Gene Autry Western Heritage Museum, Los Angeles, California

graph of *American Progress* (Fig. 43). It has a similar female figure moving across the sky from the East to the West. Shorn of wings, and secularized, Pratt's angel was transformed by Gast into a symbol for the promotion of western tourism in a book by George Crofutt.[98] Gast also reversed Pratt's symbolism of dawn and evening; his allegorical figure moves from the morning light in the East to illuminate the receding darkness to the West, drawing along behind it pioneers and railroads.

By the time of his 1872 exhibition Pratt was fifty-nine years old. Aside from one final notice in 1874 in the *Boston Evening Transcript* about how he employed photography to produce his Yosemite paintings, Pratt was not mentioned again until his death in 1880. By the end of the 1860s Pratt's work was passé—his reliance on allegory was at variance the emergence of Impressionism. Pratt's reputation was so obscure, in fact, that Henry T. Tuckerman, known as the "American Vasari," did not even mention him in his 1867 *Book of the Artist*, which is often considered the definitive accounting of American artists of the period. Pratt's obituary, entitled "An Old Artist Gone" by the Boston landscape painter Samuel L. Gerry, extolled his virtue and industriousness. It stated, "He was a contemporary with a generation of artists of which scarcely any remain ... Though Mr. Pratt was by profession a portrait painter, his feeling for landscape led him to pursue at intervals that branch of art, which had he given himself to it unreservedly, would have probably given him a high place on the roll of fame."[99] Pratt's long friendship with Cole was noted. Gerry concluded that Pratt's esteem had "ingrained goodness worth more than fame, worldly honor or pecuniary success."

The Legacy of the Pioneers of the Palette

Both borders and frontiers begin as cultural attitudes and assumptions, and reflect the intersection of conceptual landscape constructions with physical features perceived as places that demark or establish differences. Borders are not only physical and geographical boundaries, they are also sites marking national, cultural, ethnic, and racial distinctions. These distinctions do not usually emerge at once, but rather develop gradually over time, and may ultimately come to separate one people or one nation from another as the border between the United States and Mexico does today. The demarcation of borders by surveyed lines is unique to modern nation states. The U.S.-Mexico Boundary Survey of the mid-19th century marked a border that was initially very permeable, and along most of its route virtually undetectable. Today the border between the United States and Mexico is marked in places by walls of steel as impenetrable as the walls that once separated European nations during the Cold War. The U.S.-Mexico Boundary Survey led by Bartlett established much of the route of this geographical border and germinated the national, ethnic, racial, and cultural problems that confront United States and Mexico today. Instead of resolving these issues, the border first drawn by Bartlett and García Condé seems to acerbate them.

The U.S.-Mexico Boundary Survey was part of a much larger program of expansion that characterized 19th century American and European societies. Around the globe explorers were sent out from metropolitan market centers to explore, conquer, and survey landscapes never before seen by Europeans. Bartlett's survey of the U.S.-Mexico Boundary was but one chapter in this larger process. The images produced by Bartlett and Pratt are prime artifacts from this period of American expansion. Their art and literature promoted a vision of the southwestern deserts as a grand stage guarded by angels upon which would be enacted a national morality play of American progress and transcontinental expansion. In their work the region's potential for agricultural development first became evident, and its enormous supply of natural resources and suitability for an all-season transcontinental railroad were celebrated. Its scenic potential was represented in their art as wonder and spectacle that would soon attract tourists.

Bartlett's and Pratt's work engaged the ideology of Manifest Destiny by relying on established conventions for representing "history," thus providing additional justification for further expansion. With this came further appropriations of territory from Native Americans, and increasing exploitation of natural resources. The Southwest was idealized as exotic and primitive but also as highly desirable. Because the U.S.-Mexico Boundary Survey, especially Bartlett's and Pratt's work, the process began of imprinting on the American antebellum imagination a vision of the Southwest as a strange and wonderful region that was full of commercial potential. In doing this the Boundary Survey made a lasting contribution to the defining of the "myth of the West."

Of primary importance was the contribution of surveying a plausible route for a railroad west to the Pacific Ocean. Ironically, not once, but twice, Bartlett's "failure" was rewarded: The ruckus caused by his "mis-drawing" of the border led to a second survey by Major Emory that completed Bartlett's work and mapped the Gadsden Purchase. Between Bartlett and Emory the borderlands began to gain a more focused cultural definition, complete with strange desert landscapes, exotic plants, animals, and native inhabitants. The inept way that Bartlett's survey functioned caused the government to assign later western surveys to a "professional" cadre of military (Army) topographers. The advantages of bringing artists along on such expeditions was clearly established and in later expeditions their role was strengthened.

Bartlett and Pratt also refined the cultural usefulness of collaborations between artists and explorers. Their relationship was typical of later associations between artists and scientists: Albert Bierstadt with Thomas Starr King at Yosemite, and Thomas Moran with Ferdinand V. Hayden at Yellowstone. These later artist-explorers amplified the value of the collaboration established by Bartlett and Pratt in the early 1850s. Bartlett and Pratt were unusual to the degree that Bartlett was an accomplished and productive artist whose talent was comparable to Pratt's. Through the popular success of their cultural productions—the books, illustrated reports, drawings, prints, and paintings—the American Southwest was opened visually and conceptually for the first time to Easterners, and was done in a manner that made it acceptable to established elites in the Northeast.

By the end of the century the contributions of Bartlett and Pratt were forgotten. Not until the mid-1960s, when historian Robert Hine first excavated, published, and exhibited their work, would their historical importance be recognized. Now, almost three decades later, the time has come for another look at these artists who, "led by an unseen hand," were "teachers and coworkers with the pioneer, the man of science, and the soldier" in the colonization of "this mighty continent."[100]

NOTES

Drawing Borders: Art and the Cultural Politics of the U.S.-Mexico Boundary Survey 1850-1853

1. The most comprehensive monograph on the Bartlett survey remains, Robert V. Hine, *Bartlett's West: Drawing the Mexican Boundary* (New Haven: Yale University Press, 1968). Hereafter cited as Hine, *Bartlett*. Also useful is Odie B. Faulk, "John Russell Bartlett and the Southwest: An Introduction," in John Russell Bartlett, *Personal Narrative of Explorations and Incidents in Texas, New Mexico, California, Sonora, and Chihuahua*, 2 Vols. (Chicago: Rio Grande Press, 1965, orig. pub. 1855), hereafter cited as Bartlett, *Narrative*. For an overview of exploration in the West see William H. Goetzmann, *Army Exploration in the American West, 1803-1863* (New Haven: Yale University Press, 1959), 153-208.

2. "Speech of the Hon. V. E. Howard of Texas, in the House of Representatives, July 6, 1852," *Appendix to the Congressional Globe, XVV*, 776-781.

3. Howard R. Lamar, *The Far Southwest, 1846-1912: A Territorial History* (New Haven: Yale University Press, 1966), 92.

4. Lamar, *The Far Southwest*, 92.

5. no title, *The [New York] Evening Post*, 23 September (1851), 1. The report from an unnamed correspondent accompanying the boundary survey stated: "We understand that the business of the Commission to adjust the boundary between Mexico and United States was almost suspended in consequence of a disagreement between the Commissioner and the Surveyor on the point of commencement.... The Commission is at a dead halt. Mr. Bartlett is at war with Colonel Graham, the astronomer to the Commission." For a modern assessment of this problem see Faulk and Hine.

6. See William H. Truettner, ed. *The West as America: Reinterpreting Images of the Frontier, 1820-1920*. exh. cat. (Washington, D. C.: Smithsonian Institution Press, 1991).

7. See William Cronon, "Telling Tales on Canvas," in *Discovered Lands, Invented Pasts*. exh. cat. (New Haven: Yale University Press, 1992), 44.

8. See J. Gray Sweeney, "The Artist-Explorers of the American West, 1860-1880," unpublished Ph.D. dissertation, Indiana University, 1975.

9. See Susan Danly and Leo Marx, *The Railroad in American Art: Representations of Technological Change* (Cambridge: M.I.T. Press, 1988); Sandra D'Emilio and Suzan Campbell, *Visions and Visionaries: The Art & Artists of the Santa Fe Railway* (Salt Lake City: Peregrine Smith Books, 1991); T. C. Mcluhan, *Dream Tracks: The Railroad and the American Indian, 1890-1930* (New York: Harry N. Abrams, 1985). See also Dona Brown, *Inventing New England: Regional Tourism and the Nineteenth Century*. (Washington, D.C.: Smithsonian Institution Press, 1995).

10. Raymond Williams, *The Country and the City* (New York: Oxford University Press, 1973), 121, 128.

11. See Goetzmann's *Army Exploration*, and William H. Goetzmann, *Exploration and Empire* (New York: Knopf, 1967).

12. See "Report of the Secretary of the Interior," *The* [Boston] *Daily Courier*, 15 December (1852), 1.

13. Ron Tyler, *Prints of the West* (Golden: Fulcrum Publishing, 1994). Particularly useful is Chapter 4, "The Discovery of the West: Government Patronage," 69-108.

14. William H. Gerdts and James Yarnall, *Index to American Art Exhibition Catalogues* (Boston: G. K. Hall, 1986), 192, #s 3800-3802. Hereafter cited as Gerdts and Yarnall, Index. The subjects for these drawings were diverse: a still life of a cake, a St. Bernard dog, and a seascape.

15. John Russell Bartlett, "Autobiography," unpublished ms. John Carter Brown Library, Brown University, 21.

16. Robert V. Hine, *Bartlett's West: Drawing the Mexican Boundary* (New Haven: Yale University Press, 1968), 6-7.

17. Bartlett, "Autobiography," unpublished ms. John Carter Brown Library, Brown University, 50-51.

18. Thomas Cole, "Lecture on American Scenery," in Marshall Tymn, ed. *The Collected Essays and Prose Sketches* (St. Paul, MN: John Colet Press, 1980), 200.

19. "Labors of Artists," *Boston Transcript*, 15 June (1848), 2.

20. For an account of the cultural and political agendas underlying the legend of Cole's "discovery" by artists John Trumbull, Asher B. Durand, and William Dunlap see Alan Wallach, "Thomas Cole: Landscape and the Course of American Empire," in *Thomas Cole: Landscape into History*, exh. cat. William H. Truettner and Alan Wallach, eds. (New Haven: Yale University Press, 1994), 23-28.

21. William Dunlap, *History of the Rise and Progress of the Arts of Design in the United States,* Vol. 3 (Boston: C. E. Goodspeed, 1918; orig. pub. 1834), 171. See also Catherine H. Campbell, "Two's Company: The Diaries of Thomas Cole and Henry Cheever Pratt on Their Walk through Crawford's Notch, 1828," *Historical New Hampshire,* Winter 1978, 309-33.

22. For a discussion of this visit see Pamela Belanger, "Infinite Treasures in a Small Room: Artists and Tourists at Mount Desert Island," unpublished M.A. thesis, Arizona State University, 1992. For the reattribution of *Moses on the Mount* see Ellwood C. Parry, *The Art of Thomas Cole: Ambition and Imagination* (Newark University of Deleware Press, 1988), 86-87.

23. Gerdts and Yarnall, *Index,* #72585, 2846.

24. Gerdts and Yarnall, *Index,* #72581, 2846.

25. Henry Cheever Pratt to J. R. Bartlett, 23 July 1851, Bartlett Collection, John Carter Brown Library.

26. Cited in "Cole's America: An Introduction," by Christine Stansell and Sean Wilentz, in William H. Truettner and Alan Wallach, eds. *Thomas Cole: Landscape into History,* exh. cat. (New Haven: Yale University Press, 1994),18.

27. Bartlett, *Narrative,* I:303.

28. See John Barrell, *The Dark Side of the Landscape: The Rural Poor in English Painting 1730-1840* (Cambridge: Cambridge University Press, 1980).

29. Bartlett, *Narrative,* I:118.

30. Bartlett, *Narrative,* I:119.

31. Bartlett, *Narrative,* I:247-48.

32. Bartlett, *Narrative,* II:192. "A number of fish called by Major Emory, 'Gila trout' were caught near our camp by Mr. Pratt with a hook and line. They proved very palatable, where fresh fish is such a rarity."

33. Bartlett, *Narrative,* II:106.

34. Bartlett, *Narrative,* I:206. This marker has been destroyed, and the bottle containing the signed documents has apparently been lost or taken.

35. Bartlett, *Narrative,* I:156.

36. Bartlett, *Narrative,* I:160.

37. Bartlett, *Narrative,* I:183.

38. Bartlett, *Narrative,* I:193.

39. Samuel Magoffin to James W. Magoffin, Esq. St. Louis, 19 January 1853, in "James Magoffin," typescript manuscript copy at Magoffin Historic Site, no date. n. p. Samuel's wife Susan Shelby Magoffin wrote the now famous diaries of her trip down the Santa Fe Trail in 1846-1847. See *Down the Santa Fe Trail and Into Mexico: The Diary of Susan Shelby Magoffin 1846-1847.* ed. Stella M. Drumm (Lincoln: University of Nebraska Press, 1982).

40. Bartlett, *Narrative,* II:402.

41. For a history of El Paso, and the establishment of Coons's Ranch see: W. H. Timmons, *El Paso: A Borderlands History* (El Paso: Texas Western Press, 1990), 103-116. For Coons's relationship with Ponce see: J. J. Bowden, "The Ponce de Leon Land Grant," *Southwestern Studies* 24 (1969),3-9.

42. Pratt to Bartlett, 23 July 1851. Bartlett Archives, John Carter Brown Library, Brown University. This painting is dated 1853. It is unclear whether this is the painting Pratt produced for Coons or the view he promised to paint for Bartlett from the landscape he produced in 1851. Its provenance is from an unnamed collector in New England, and it entered the collection of the Memorial Museum in the early 1940s at about the same time as Seth Eastman's watercolors for Bartlett were donated to Brown University. The 1853 date of the picture and its New England provenance suggests that it is probably a replica of the 1851 painting Pratt completed for Coons. It may have been owned by Bartlett and descended through Bartlett's family to be eventually disposed of around the same time as Eastman's watercolors by Bartlett's descendants.

43. Bartlett, *Narrative,* I:192-93.

44. For an account of Coons's activities during the period see: Ben E. Pingenot, "The Great Wagon Train Expedition of 1850," *Southwest Historical Quarterly* October (1994), 184-225.

45. Bartlett, *Narrative,* I:303.

46. Bartlett, *Narrative,* I:307.

47. *San Diego Herald,* 14 February (1852),2.

48. Gerdts and Yarnall, *Index,* #72329, 2844.

49. Bartlett, *Narrative,* II:264-65.

50. Bartlett, *Narrative,* II:283.

51. Bartlett, *Narrative,* II:345-49. Bartlett noted, "The general character of this extensive range of buildings is the same as that of the Casas Grandes, near the Pimo [sic] villages, on the Gila, and they are unquestionably the work of the same people."

52. Bartlett, *Narrative,* II:296-7.

53. Bartlett, *Narrative,* II:295.

54. See Alan Wallach, "Making a Picture of the View From Mount Holyoke," *Bulletin of the Detroit*

Institute of Arts 66,1 (1990): 35-46; and Albert Boime, *The Magisterial Gaze: Manifest Destiny and American Landscape Painting, c. 1830-1865* (Washington, D.C.: Smithsonian Institution Press, 1991).

55. Bartlett, *Narrative*, II:98.

56. Bartlett, *Narrative*, II:298.

57. Bartlett, *Narrative*, II:299.

58. Bartlett, *Narrative*, II:299-300.

59. Bartlett, *Narrative*, II:296.

60. Bartlett, *Narrative*, I:412-13.

61. Bartlett, *Narrative*, I:414-15.

62. Bartlett, *Narrative*, II:307.

63. Bartlett, *Narrative*, I:423-24.

64. Bartlett, *Narrative*, I:195.

65. Bartlett, *Narrative*, II:394.

66. Henry Cheever Pratt, *Catalogue of Original Paintings of California and other American Scenery now on Exhibition and for Sale at the Art Gallery of the "Studio Building"* (Boston: W. F. Brown, 1872), #6. Hereafter cited as Pratt, Catalogue.

67. Bartlett, *Narrative*, II:197-98.

68. Bartlett, *Narrative*, II:205.

69. Bartlett, *Narrative*, I:224.

70. Bartlett, *Narrative*, I:iv.

71. Bartlett, *Narrative*, I:197.

72. Bartlett, *Narrative*, I:234.

73. Bartlett, *Narrative*, I:230.

74. Pratt, *Catalogue*, #12.

75. Bartlett, *Narrative*, II:590. Appendix B. *Official Instructions from the Department of the Interior to John R. Bartlett, Commissioner on the part of the United States, for running and marking the Boundary Line between the United States and the Republic of Mexico*, August 1, 1850.

76. Bartlett, *Narrative*, II:66.

77. Bartlett, *Narrative*, II:589. Appendix B

78. Bartlett, *Narrative*, II:575.

79. Bartlett Papers. "The Pacific Railroad," New York, 21 December 1853 (no publisher cited). John Carter Brown Library, Brown University.

80. Bartlett, *Narrative*, I:x.

81. Bartlett, *Narrative*, xii.

82. Bartlett Papers, "Bartlett's Narrative," *Boston Post*, 18 July 1854, no page cited. John Carter Brown Library, Brown University.

83. From the *Boston Daily Advertiser*, no title or date, Bartlett Papers, 318. John Carter Brown Library, Brown University.

84. *Boston Daily Advertiser*, Bartlett Papers, 318.

85. *Boston Daily Advertiser*, Bartlett Papers, 352.

86. This trip to California and the date of 1870 can be deduced from Pratt's 1872 exhibition catalog. It contained a painting entitled, *A Near View of Hutchins Hotel As It Was in 1870*. Evidently by 15 March 1871 Pratt was back in Boston because a notice, "Art Items," in the Boston Evening Transcript, comments on his "Yo Semite" pictures, 2.

87. Sigma, "Works of Art," Boston Evening Transcript, 5 December (1855), 1. The two earlier and shorter reviews are: Sigma, "Mr. Henry C. Pratt's Pictures," *Boston Evening Transcript*, 7 December (1855), 2; and Sigma, "Works of Art," Boston Evening Transcript, 21 November (1855), 2. According to Merl M. Moore, "Sigma" may be one W. H. Domitt, about whom little is known, except that he reviewed for the *Transcript* in the 1850s.

88. Sigma, "Works of Art," *Boston Evening Transcript*, 5 December (1855), 1.

89. Sigma, "Works of Art."

90. Bartlett, *Narrative*, II:209.

91. Bartlett, *Narrative*, II:238.

92. This later title is certainly the same picture as it was retitled for its exhibition at the National Academy of Design annual exhibition. In his 1872 exhibition in Boston Pratt added the note, "It is the *Cereus Giganteus*, sometimes growing to fifty feet in height, with numerous branches. The Fruit is sweet, resembling figs." Pratt, Catalogue. See also *National Academy of Design Exhibition Record* (New York: New York Historical Society, 1943), 80.

93. See, "A Peculiar Cactus," *Boston Evening Transcript*, 22 May (1872), 1; and a one sentence note in the same paper for 23 May (1872) informing readers that Pratt's exhibition would close the next day, 2.

94. Bartlett, *Narrative*, I:278.

95. Bartlett, *Narrative*, II:286-87.

96. Pratt, *Catalogue*, # 5, #13.

97. *Boston Evening Transcript*, 24 May (1862), 4.

98. George A. Crofutt, Crofutt's New Overland Tourist and Pacific Coast Guide. (Chicago: Overland, 1878-1879).

99. S.L.G. [Samuel Lancaster Gerry], "An Old Artist Gone," *Boston Evening Transcript*, 30 November (1880), 4.

100. Samuel G. W. Benjamin, "A Pioneer of the Palette" Thomas Moran, *Magazine Art*, Vol.V, (Feb. 1882) 89-90.

LUCRETIA HOOVER GIESE

ARTIST COLLABORATORS: A Surrogate Hand- Seth Eastman's for Bartlett's

Late in 1849, U.S. Army Captain Seth Eastman arrived in Washington, D.C. on leave. He had finally secured what he anticipated being a significant professional assignment that would be helpful in terms of remuneration. He would illustrate Henry Rowe Schoolcraft's six-volume *History of the Indian Tribes of the United States*, one of the period's grand "scientific" projects underwritten by the U.S. government. Neither Schoolcraft's nor Eastman's passage through the turbulent waters of Washington in the 1850s would be easy or conclude as each had hoped. Eastman's connections there, however, did lead him to another "patron," John Russell Bartlett, Commissioner of the U.S.-Mexico Boundary Survey.[1]

Seth Eastman, an easterner by birth and upbringing, chose a career path that took him far from his native Maine.[2] He entered the U.S. Military Academy at West Point at the age of sixteen, and from 1824 until his retirement in 1867, the military framed his life. His first posting in 1829 took him to Fort Crawford, Wisconsin, then to Fort Snelling, Minnesota, and on to Louisiana briefly in 1832. After seven years at West Point as an instructor, he served in Florida, again at Fort Snelling from 1841 to 1848, and then in Texas before being granted leave. A return assignment in Texas and other stations between 1855 and 1863 ended Eastman's active military career, although he was brevetted a brigadier general in 1866. These postings to which he submitted, with great frustration on occasion—his wife described him as "buried alive in Texas" in 1855—nevertheless brought him in contact with the material of his life's work, the Native American.[3]

Eastman's avocation was art, and the Army provided his artistic training. He studied topographical drafting at West Point and, while an assistant drawing instructor there from 1833 to 1840, published a *Treatise on Topographical Drawing* in 1837. He studied painting under Robert W. Weir, a fellow instructor at West Point, whose collection of prints and books on art Eastman also found helpful. He was offered, but declined, a professorship of topographical drawing and painting at another institution. Significant fuel for his artistic aspirations came when his work was accepted for the 1836 National Academy of Design Annual, and honorary membership was granted to him in 1838. Eastman also considered collaborating on a panorama depicting the Mississippi, illustrating a book about the river, and assembling an Indian gallery of paintings.[4] He did manage to sell to the American Art-Union. This organization acquired paintings for its lottery and produced engravings from them for distribution to its sizable membership. Clearly Eastman took himself seriously as an artist, keeping up his membership in the National Academy of Design until 1860.

The most significant ingredient in Eastman's personal life was Mary Henderson, the daughter of an Army surgeon stationed at West Point, whom he married in 1835. An educated woman, Mary Eastman wrote about the Native Americans who inhabited areas where Eastman was posted. The first and best known of these publications was *Dahcotah; or, Life and Legends of the Sioux around Fort Snelling.*[5] Her husband's illustrations often augmented her narratives. She was ambitious for her husband and worked hard on his and the family's behalf, not hesitating to opportune any influential person with whom the Eastmans had contact. After Eastman's second Fort Snelling assignment, she stayed behind in the east to raise their five children and help propel her husband's career, which she did adroitly, even aggressively. Correspondence shows her actively trying to arrange leaves for Eastman or transfers to the east. They both sought a posting to Washington, D.C. where preparations were underway for Schoolcraft's Indian history within the Department of Interior. Late in 1849, Eastman arrived in Washington, where he remained for five years while working for Schoolcraft.

Bartlett was in Washington himself by 1853, after having been dismissed as Commissioner and now pushing for publication of a report of the boundary survey comparable to Schoolcraft's. Bartlett had known Schoolcraft since the 1840s and had discussed Schoolcraft's project with him. Both understood the value of illustrations. Even before Bartlett's appointment as Commissioner of the U.S.-Mexico Boundary Survey, Schoolcraft had written informing Bartlett that Eastman "has been detailed to furnish illustrations" for his Indian history.[6] Unhappily for Bartlett, a resolution before Congress in the spring of 1853 calling for a report corresponding "in style and form" to Schoolcraft's went down to defeat.[7] But that August, Bartlett, remembering Schoolcraft's enthusiastic comments about Eastman, was in touch with the artist.[8] Henry Cheever Pratt and Bartlett took quite adept renderings on the survey, but evidently needing an image of San Antonio, Bartlett approached Eastman, who had not accompanied the U.S.-Mexico Boundary Survey.[9] Eastman, suffering from diarrhea, responded that he did not know "when I shall be well enough to make a good drawing for you, from my sketch of the Mission of Conception," so instead "cut the original sketch from my sketch book ... made on the spot on the 28th Nov. 1848."[10]

It is easy to understand why Bartlett sought out Eastman. He was working for Schoolcraft, by then Bartlett's rival for government funding, and like Bartlett, was caught up in what Brian Dippie has called "the politics of patronage."[11] Eastman came with Schoolcraft's imprimatur in addition to his own impressive credentials as a topographical draftsman and artist, and was familiar with at least the Texas portion of the survey's territory, having been stationed there. The quality of the San Antonio sketch which Eastman gave Bartlett, to say nothing of Eastman's generous accommodation of Bartlett's request, may have furthered Bartlett's interest.

Eastman completed at least thirteen watercolors for Bartlett, a number confirmed by Eastman himself. Writing from "the wilderness" of Fort Chadbourne, Texas, in October 1855 after reassignment, Eastman bluntly prodded Bartlett to "forward to my wife in Washington a draft for the amount due me for the thirteen sketches."[12] The month before, Mary Eastman had broached Bartlett with the matter of Eastman's "call ... about some money.... He wishes me to say to you that if you desire to have two more drawings that he made for you, that they were left, finished, with me."[13] Although there is discrepancy here about the total number Eastman made for Bartlett, Eastman appears to have held Bartlett responsible for only thirteen. Bartlett probably had these in 1855—Eastman does not mention sending them—and presumably paid the artist since subsequent correspondence between Bartlett and the Eastmans remained cordial.[14]

What these letters document is the Eastmans' pecuniary concerns.[15] Again and again, Mary Eastman complained to a variety of respondents that an army officer's pay did not cover their needs; compensation above that for Eastman's work as illustrator had not been possible for them (or Schoolcraft, so he claimed) to secure.[16] In addition, it rankled that Schoolcraft asserted ownership of Eastman's plates and reissued them in a privately published edition of the Indian history, even though at the same time Eastman reused some in his wife's *Chicóra, and Other Regions of the Conquerors and Conquered* and other works, presumably undertaken for financial gain. It may be that a measure of spite against Schoolcraft came into play, but Eastman almost certainly took on Bartlett's request for monetary reasons.

Eastman's involvement with Bartlett grew beyond the simple Texas sketch initially requested. Eventually, Eastman's watercolors spanned the years of Bartlett's campaign and the territory he covered, evidently a conscious decision on Bartlett's part. Elaborate inscriptions penciled on the verso of the watercolors indicate what is rendered.[17] Bartlett's text and illustrations in *Personal Narrative of Explorations and Incidents in Texas, New Mexico, California, Sonora, and Chihuhua. Connected with the United States and Mexican Boundary Commission* substantiate this material, and it is further verified by annotated on-site drawings by Bartlett or Pratt. For the sake of continuity, the inscriptions are used here as titles.

The Eastman Watercolors

The "first" of the watercolors is *Fording Devil's River, Texas* (Pl. 26). Bartlett's route to El Paso led northwest to the Pecos River, the "largest tributary of the Rio Grande."[19] Before that point, the party crossed the Devil's River, or the San Pedro, several times. Their "Camp near the last crossing of the Devil's River [was] 230 1/2 miles from San Antonio."[20] At the end of October 1850, Bartlett and his crew navigated the Pecos at Horse-head Crossing at some risk to mules and men. Bartlett chose to illustrate the later more dramatic episode described at length in *Personal Narrative* with his drawing "Crossing the Pecos."[21] Bartlett's image foregrounds a covered wagon and mule team floundering in deep water as men on horseback attempt to urge the mules forward. In contrast, Eastman's watercolor shows men and animals as insignificant in the vast open landscape, fording a river without difficulty as they are observed by two Indians on a high rocky outcropping to the left. In this case, no known Bartlett sketch corresponds to Eastman's watercolor, nor does Bartlett describe Devil's River in *Personal Narrative.*

After months at El Paso, in April 1851, Bartlett established his headquarters at the abandoned copper mines of Santa Rita del Cobre; that October, after the commission left, the site became Fort Webster.[22] Bartlett described it as a valley "six thousand two hundred and fifty feet above the level of the sea" from which the eight thousand-foot Santa Rita Mountain forms "the beginning of a range of bold, rocky bluffs of trap, of a grayish hue, which extend some twenty miles to the south, and gradually drop off into the plain." The mines themselves, worked for about forty years until 1838, consisted of over fifty adobe structures including a "Presidio" in good condition and others in ruin.[23] This general area is the subject of three of Eastman's watercolors.

In Eastman's *Cañon leading to the Copper Mines, Santa Rita del Cobra* [sic] (Pl. 27), a mountainous ridge stretches left to right with a river winding into the distance along the valley floor. The distinctive flat-topped peak to the left in the watercolor appears in the center of "Cañon leading to the Copper Mines, from the South" in *Personal Narrative.*[24] Bartlett started but did not finish this formation in one drawing; in another, he placed the mountains lower and closer to the viewer than does Eastman, and added two figures and a profusion of trees to the foreground.[25]

Santa Rita del Cobra [sic] (Copper Mines) *Rocky Mountains New Mexico* (Pl. 28) shows an encampment of tents to the left with mine buildings spread across the valley surrounded by gently sloping hills rising to distant mountains. Overall, the scene resembles "Valley of the Copper Mines from the South" in *Personal Narrative.*"[26] But Eastman's panoramic view conforms even more closely to a drawing inscribed "OSK Bessau from a sketch by J. R. Bartlett."[27] In this drawing there are two figures in the foreground, one carrying a gun. Perhaps Bartlett intended a personal reference, as he often hunted (and sketched) when away from camp.

Extensive adobe ruins dominate the high foreground of *Bufa del Cobra* [sic] *High Bluff, Rocky Mts 1800 ft above the valley + 8000 ft above the level of the sea* (Pl. 21).[28] Covered wagons stand nearby, and men rest under a large tree to the right with tents and horses beyond. A peak with a nearly vertical face is prominent in the background. Although this formation also appears in "Presidio at the Copper Mines" in *Personal Narrative*, the fort and other well-preserved structures are the wood-cut's focus.[29] No initial drawings by Bartlett have surfaced.

In May 1852, Bartlett and a party of men traveled into Sonora as far south as the town of Arispe. They would be in this area again, reaching Magdalena in early October. *Cañon leading to Magdalena Sonora, Mexico* (Pl. 20) shows their approach along a river bordered on either side by high cliffs opening onto a view of foothills backed by high mountains, the distinctive cactus *petahaya* or "giant candelabra," as Bartlett called it, which he noted first seeing at Magdalena, rising from the top of the right-hand canyon wall.[30] Eastman's watercolor corresponds not only to Bartlett's drawing inscribed "Cañon between Bacuachi and Arispe" but to his own drawing inscribed along the bottom "Cañon between [word omitted] + Magdalena" and across the top, "Painted To be painted."[31] Indeed it was, with Eastman adding but a few more figures to the train in the foreground.

Bartlett described encountering, "soon after leaving Bacuachi," an "extraordinary defile" whose "perpendicular sides ... rose six or eight hundred feet above on either hand" taking the shape of "Columns, turrets, towers and pyramids ... decked with brilliant flowers or bearing strange cacti, ..."[32] The location seems further secured by a notation "Columnar rocks, near Arispe" on a list of "Subjects for illustrations" Bartlett compiled.[33] Yet the proximity of Arispe to Magdalena and other cañons in the area may have caused even Bartlett some confusion. His drawing "Magdalena, Sonora" appears to have both a June (partly obscured) and a 3 October 1851 date.[34] A later oil by Pratt recapitulates Bartlett's composition and that of Eastman's drawing/watercolor, although the foreground rocks and vegetation, figural elements, and formation of the cliffs are simplified, and Pratt added more plant life to their rims.[35]

Following their return from Sonora, Bartlett and his team finally left the copper mines for Mexico. *View from Camp near the Boundary line 4th day from Fort Webster* (Pl. 29) depicts the team's encampment a few days journey from headquarters. High mountainous terrain in Eastman's watercolor serves as backdrop to an arid landscape supporting some of the region's principal shrubs such as mesquite chaparral and species of yucca that Bartlett enumerated. Sketches for this watercolor are unlocated.

On 30 August 1851, their guide pointed out a "lofty and continuons [sic] range known as the *Sierra Chiricahui*" dominated by "*El Pitoncillo* [sic], or Sugar Loaf, a mountain of this form in the high range alluded to, with an opening near it where he said General Condé [the Mexican Commissioner] was encamped."[36] Two days later Bartlett wrote in his journal that he "took my sketch book with my servant ... and scouted around in search of the picturesque ... took two sketches; one giving a view of the pallisado mountain as we called it with the plains on the right, and the mountains on the Gila [River] in the distance." He made a summary rendering of mountains on the preceding page.[37] Eastman's watercolor *Pass at the Pitoncillo* [sic] (Pl. 30) depicts the same mountains seen from the pass through which a caravan of men and pack animals make their way down toward a broad plain far below.[38] A more complete drawing by Bartlett dated 3 September, shows the same range from the valley floor with men and tents in the foreground. Bartlett's third drawing of the day before provided Eastman with the template for his watercolor.[39] The range is backdrop to a marsh in still one more drawing dated September 3 and inscribed "5th camp from the Cobra [sic]."[40] Two wood-cuts in *Personal Narrative* are relevant: "Chiricahui Mountains" shown from the valley; and "La Punta de Sauz Cienega [sic]," the tip of the marshy ground before the mountains.[41]

Later in September, some members of the commission separated from the team in order to begin surveying the Gila River. Lieutenant A. W. Whipple, U.S. Topographical Engineers, in charge of the astronomical and magnetic surveying undertaken, described certain difficulties encountered "below the junction of Rio San Francisco." At what he called "the entrance to the cañon of the Pinal Lleño [sic] mountains," the canyon walls "rose perpendicularly from five to fifteen hundred feet in height" forcing the party to climb them or wade the river.[42] Bartlett's sketch of this precipitous river gorge is dated 27 October 1851.[43] It is also inscribed "sketch for Frank Wheaton." Perhaps Wheaton, one of Whipple's assistants, later described the scene to Bartlett, which Whipple's report substantiates they reached on 27 October. To Bartlett's image, which Eastman's watercolor *Great Cañon, Rio Gila* (Pl. 31) matches almost exactly, Eastman added a rare comment on the operations of the commission. Two men on a narrow rocky shelf dwarfed by the steep right-hand cañon wall appear to be carrying out surveying operations.[44]

While Whipple was on the Gila River, Bartlett took his group south to the town of Santa Cruz. There on 23 September 1851, after crossing "desolate and barren plains," they found "luxuriant herbage,... and live oaks." Delighted with the almost park-like vista as "the train was passing down the mountain, [Bartlett] ... stopped with Mr. Pratt to enjoy the scene, which he hastily transferred to his sketch-book."[45] Pratt's preliminary drawing, dated 23 September, as well as his watercolor reproduced full-page as "Valley leading to Santa Cruz, Sonora" in *Personal Narrative* have survived.[46] Eastman's *Santa Cruz Valley* (Pl. 32) conforms to these, though not slavishly. In both, the valley stretches below a hillside sloping toward the left with mountains across the distance. Eastman, however, omitted the corresponding, more gently rolling ground to the left where Pratt placed a mule-drawn wagon, brought the right-hand hillside further forward, and inserted the figure of a man seated beneath one of two large trees, sketch- or notebook in hand. Bartlett may have suggested this comment on his or Pratt's activities, or Eastman may have picked it up from reading *Personal Narrative.* A view of the town itself for which Bartlett made a drawing appears as "Santa Cruz, Sonora" in *Personal Narrative.*[47]

More than six months later, having sailed to San Diego from Acapulco, Mexico, visited the Napa Valley, and returned to San Diego, Bartlett began his trek back to the Gila River and eventually to El Paso. On 2 June 1852, Bartlett stopped at *Vallecito, California* (Pl. 33), about eighteen miles from San Felipe. The name *Vallecito,* as we are told in *Personal Narrative,* refers to "a little valley, surrounded by lofty and barren mountains" where soldiers and provisions for Fort Yuma on the California border were housed. Eastman's watercolor of Bartlett's campsite accentuates the tents, mules, and wagons set against mountains that Bartlett wrote about, but does not include what he called willow bushes among which "we pitched our tents."[48] No drawings are known.[49]

Later that month, the party reached the Pima and Maricopas Indian villages near the confluence of the Salinas and Gila Rivers; their location is clearly marked on a map made for the boundary survey.[50] Sketches by Bartlett and Pratt gave Eastman ample material to work with for his *Village of the Pimo* [sic] *Indians Gila River* (Pl. 8).[51] But Eastman had been introduced to their culture somewhat earlier by "drawings of pottery found on the Gila river" and a Pima vocabulary sent to Schoolcraft by Major Emory who was attached to the boundary survey.[52] Eastman's watercolor focuses on Indian life with the distinctive Chiricahui Mountains as backdrop. One

Indian wears a lozenge-shaped armature on the back, possibly one described by Bartlett; another weaves at a loom staked out on the ground, while others go about indefinable activities.[53] Two drawings by Eastman presumably after unlocated Bartlett originals depicting an Indian weaving and Indian houses, relate to "Indian weaving" and "Pimo [sic] man and woman" in *Personal Narrative*.[54] Pratt, too, whose sketch is very close to Eastman's watercolor, made the villagers, their dwellings and animal pens prominent; it is the illustration "Coco-Maricopa Village" in *Personal Narrative*.[55] Another wood-cut in *Personal Narrative* "Cultivated Fields and Villages of the Pimo [sic] Indians," however, based on a drawing by Bartlett, distances the Indians and their village altogether.[56]

On the high table land near El Paso, Bartlett encountered many storms. One occurred 13 August 1852. Near Rio Santa Maria, on a "level plain" before mountains "1500 to 2000 feet high, ... resembling castles, domes, and minarets ... isolated, or in groups," ominous black clouds appeared. Hurriedly Bartlett's party set up camp, completing their task just as the storm hit. Even so, "but for the extra ropes and stays put to the tents, all would have been blown away."[57] In Eastman's *Camping in a Storm on the Mexican Plateau* (Pl. 34), men struggle against wind and rain advancing from the left as they work to establish camp set before mountains with what Bartlett described as "fantastic outlines." A drawing by Bartlett, dated 15 August 1852, of the "Mountains on Correlitas Road 40 miles from El Paso" confirms Eastman's image, even to rain at the left indicated by a series of vertical lines marked "Thunderstorm moving up."[58] There was rain on 15 August, as well as the days before, according to *Personal Narrative*. The mountains reappear in a Bartlett drawing dated "Oct 16, 1852" inscribed "... looking N. E. toward Rio Grande march from El Paso to Chihuahua."[59] Bartlett was on his way home.

In early October, El Paso behind him, Bartlett went back over the Mexican plateau to Corpus Christi. En route his party stopped at a stream emptying into "a small pond containing a couple of acres, in which many ducks were seen," known as Lago de los Patos or Duck Lake.[60] His panoramic drawing of this fresh water reservoir, dated 14 October 1852, the day the party arrived, extends across two sketchbook pages.[61] This drawing informed Eastman's own drawing (annotated with "waterfowl," "Tents + Wagons," "Bushes 15 ft high," "steam or dust" beneath a distant peak labeled "Volcano," and so on), which he used for *Camp near Lake de los Patos, Chihuahua* (Pl. 35).[62]

In both Bartlett's and Eastman's drawings, a truncated volcanic mountain is prominent, but Eastman emphasized camp wagons and tents and gave the pond in the foreground an outlet. More significantly, he tightened Bartlett's composition, marking off several inches from the left edge and even more from the right, and wrote across the top "To be Painted" and "Painted."[63]

This grouping suggests Eastman's method of working. He probably studied Bartlett's sketches first, found today either in sketchbooks or as discrete sheets, then made his own more articulated drawings, of which only a few are known. With these in hand, he began with watercolor on a separate sheet of paper after lightly penciling in a cursory outline. Bartlett's drawings provided topographical information that Eastman was well-equipped to clarify, but rarely included color notations. When these are indicated they are limited in nomenclature and localized (for example, "purple," "blue," "dark olive," "Lt green" on one drawing), a pattern repeated in his accounts in *Personal Narrative.*[64] Eastman's palette tends toward mixtures such as reddish purples, bluish lavenders, pinks and yellow-beiges, yellow-greens, and warm, pale blues, but words are inadequate to describe the subtlety and brilliance of the color. Eastman was evidently forced to call upon his memories of Texas. The result is a certain sameness of coloration, warranted perhaps by the territory itself, that enhances and in no way detracts from the watercolors' overall effectiveness.

Eastman's artistic training and years of experience are fully displayed in these watercolors. His awareness of landscape conventions is evident in his use of framing devices, foreground figures, diagonal leads into the middle distance, and visual halts to the background; and his skill with handling watercolor is demonstrated in layered washes, holding of certain areas in reserve, and creation of highlights through selective scrapping. Such proficiency made it possible for Eastman to convey on a sheet of paper approximately 10 x 12 inches the borderland's concrete topographical features and its vastness, light, and atmosphere. Because of his artistry, the watercolors appear wonderfully and utterly unrehearsed. Bartlett undoubtedly recognized that Eastman's work surpassed his initial sketchings.

It is noticeable, however, that Eastman's watercolors tell us little about the operations of the survey.[65] Men lead or ride animals through the landscape, set up tents, rest beneath trees, but only once engage (apparently) in actual surveying. In any case, figuration is small and has little narrative purpose. Instead, Eastman's watercolors celebrate the landscape itself. That is logical. In the end perhaps all Bartlett could claim to have accomplished were the thousands of miles traveled and actual sites seen. For this reason above all others, Eastman was the obvious choice of illustrator for Bartlett's *Personal Narrative.*

The watercolors also tell us about both Bartlett's and Eastman's ambitions, and their different attempts to work the Washington political scene in the mid-19th century. To both, the government seemed the best and possibly the only patron for their purposes. As a military officer, Eastman looked to the government for consistent employment and professional status; it provided that. It also opened up important illustrative work in government-sponsored publications. For his part, Bartlett, a bibliophile, believed reasonably (though incorrectly as it turned out) that only the government could and should absorb the cost of publishing material that it had commissioned, including of course his own report. Both Bartlett and Eastman experienced the effects of congressional politics and fickleness, and suffered disappointment.[66]

Even so, Bartlett might have used Eastman's remarkable watercolors after all; but he did not. The reasons were financial considerations, perhaps a hint of parsimony, and Bartlett's sense of urgency. Anxious to clear his name after the debacle of dismissal as Boundary Commissioner, Bartlett wanted to publish his side of the story as quickly as possible. William W. Turner, Bartlett's friend who doubled as his editor, puzzled over Bartlett's decision to "curtail the attractions of your book in the way of illustrations because it cannot be ready by the 1st of January." He considered a change in plans unnecessary and unfortunate because "Your work ... is rather one of permanent interest + value, which will attract a due degree of attention whenever it appears, + for which a steady sale may be anticipated for some time to come."[67] But Bartlett evidently felt he did not have the time; he certainly did not have the money to produce the report as he wished.

Years before, Bartlett had labeled Schoolcraft's report "a National work" (a phrase Schoolcraft picked up) and cautioned him to hold out for quality paper and quarto size sheets for the illustrations.[68] But Bartlett, later faced with publishing privately, was concerned about cost. A block on government funding for Bartlett's U.S.-Mexico Boundary Survey in the fiscal year 1852-1853 placed him in the terrible situation of being unable to pay the survey team's way home, and there was wrangling over money for years to come.[69] Even Turner complained delicately though exasperatedly about Bartlett's neglect to compensate him for three years of editorial work on *Personal Narrative*.[70] Under these circumstances, it is perhaps understandable that Bartlett still had not paid Eastman for the requested watercolors by 1855.[71] More significantly, it would have been very costly, to say nothing of time-consuming, to reproduce Eastman's work in color; only Schoolcraft's first volume (1851) had contained some color illustrations, and it had government funding.

Bartlett alluded to painful decisions he was obliged to make in his Preface to *Personal Narrative*; in the end it contained but a scant sixteen full-page sepia-toned lithographs and ninety-four vignetted full- or partial- page woodcuts in black and white. Acknowledging the contribution of Pratt, whose sketches "convey to the mind a better idea of it [the country traveled] than the most elaborate description," Bartlett "very reluctantly" determined to "omit the most important of them from the present work, as it would detract too much from their merits to reduce them to the size of an octavo page."[72] The words "present work" are significant. Cherishing "the hope that Congress will yet decide to place the whole results of the Commission before the public in a suitable manner," he may have held the Pratts and the Eastmans out for a later official and more substantive publication.[73] But Congress did not reconsider, and Eastman's watercolors were never united with Bartlett's account.

Nor did they remain "united" with Bartlett's papers, which came to The John Carter Brown Library in 1886. Bartlett's presentation copy of *Personal Narrative* to John Carter Brown in 1866 was given to the Library by a family member in 1935. The provenance of the watercolors, however, is unknown, despite searches in several directions. All that can be said is that twelve were transferred from the Library of the Rhode Island School of Design to the Museum in 1947, the thirteenth in 1948.[74] No matter how they arrived, the fact that the watercolors by Seth Eastman are today in the Museum of the Rhode Island School of Design virtually next door to Bartlett's papers in Providence would have probably pleased both Commissioner Bartlett and artist Eastman alike.

NOTES

Artist Collaborators: A Surrogate Hand-Seth Eastman's for Bartlett's

1. I wish to thank Susan Danforth of The John Carter Brown Library at Brown University, Providence, RI, for graciously making the Bartlett Papers and drawings available to me.

2. The following biographical sketch draws upon John Francis McDermott's *Seth Eastman: Pictorial Historian of the Indian* (Norman: University of Oklahoma Press, 1961), and Brian W. Dippie's *Catlin and His Contemporaries: The Politics of Patronage* (Lincoln and London: University of Nebraska Press, 1990); they remain the basic published sources of information on Seth Eastman. Sarah Boeme, who has done extensive research on Eastman, contributed to *Seth Eastman: A Portfolio of North American Indians* (Afton Historical Press, 1995).

3. Mary H. Eastman to John Russell Bartlett, from Washington, D. C., 18, September n.d., but undoubtedly 1855. John Russell Bartlett Papers, Correspondence, Courtesy of The John Carter Brown Library at Brown University, Providence, RI (JRB Papers).

4. Eastman's fascination with the Mississippi is manifested by the approximately one hundred watercolors he later sold to Henry Lewis, a St. Louis painter. Lewis then used some for his *Das illustrirte Mississippithal* published in Düsseldorf from 1854 to 1857. McDermott's *Seth Eastman's Mississippi: A Lost Portfolio Recovered* (Urbana, Chicago, London: University of Illinois Press, 1973) describes Lewis' Eastman collection. As for Eastman's so-called Indian Gallery, the Eastmans' efforts to secure his posting to Washington had prompted several influential friends in 1848 to propose that body of work to the Secretary of War; see Dippie, *Catlin and His Contemporaries*, 182-83.

5. New York: John Wiley, 1849. Other works (all published by Lippincott, Grambo & Co., of Philadelphia) by Mary H. Eastman containing Seth Eastman illustrations include: *The Iris*, an illuminated souvenir for 1852 (1851); *The American Aboriginal Portfolio* (c. 1853); *and Chicóra, and Other Regions of the Conquerors and the Conquered* (1854).

6. Schoolcraft had gloated that Eastman's "employment in this department is deemed quite an acquisition." Schoolcraft to Bartlett, 24 December 1849. JRB Papers, Correspondence.

7. Handwritten statement dated 23 March 1853. JRB Papers, Mexican Boundary Commission Correspondence, 7:61. The resolution was debated in the Senate on 5 April; see Dippie, *Catlin and His Contemporaries*, 217, 221-23.

8. Eastman and Bartlett were acquainted at least by 1850. Eastman wrote Bartlett, 29 July 1850, "Allow me to introduce to you Mr. James Houston the young gentleman that you were kind enough to appoint on your surveying party." JRB Papers, reel 2:5. And on 28 October 1851, Eastman notified Schoolcraft that he had sent Bartlett a copy of the first volume of Indian history. Henry Rowe Schoolcraft Papers, Library of Congress, reel 40: 24211 (HRS Papers).

9. On 25 August 1853, Eastman responded to Bartlett's request for a sketch (unspecified) "as soon as I am able to make a copy of it." Two days later, Eastman sent on his sketchbook drawing adding "If it will answer your purpose you can use it." Bartlett also hired Oscar Bessau, a Washington-based artist, that fall. In a letter of 14 October to Bartlett, Bartlett's friend and editor William W. Turner acknowledged having seen "such of the illustrations as Bessau has completed." These five "seemed to me very well done; but of that you are a far better judge." JRB Papers, Correspondence.

10. Eastman, "Chapel of Concepcion at San Antonio, Texas" (sketch), Peabody Museum of Anthropology and Ethnology, Harvard University, Cambridge, MA, 1 41-72 10-85; "Mission Chapel of the Conception, San Antonio" (watercolor) (illustrated in McDermott, *Seth Eastman*, pl. 62). Eastman's image appeared as "Mission of Concepcion, Texas" in Bartlett's *Personal Narrative of Explorations and Incidents in Texas, New Mexico, California, Sonora, and Chihuhua. Connected with the United States and Mexican Boundary Commission* (New York: D. Appleton & Co., 1854), I: 44 (*PN*).

11. The phrase comes from the title of Dippie's book, *Catlin and His Contemporaries: The Politics of Patronage*. Schoolcraft and Bartlett were not unscathed by the effects of personal enmity and government reluctance to support the arts; in large measure because Congress fretted over the apparently never-ending costs of official reports. With previous publication commitments continuing to absorb

funds, potential projects worried Congress.

12. Eastman went on to say that "I would not have called on you for it but my wife writes me that she is in want of money and I have none to send her." Eastman to Bartlett, 8 October 1855. JRB Papers, Correspondence.

13. Mary Eastman to Bartlett, 18 September 1855. JRB Papers, Correspondence.

14. Four months later, Mary Eastman wrote to Bartlett about a publishing matter without mentioning or reminding him of Eastman's request of 8 October 1855. She was still corresponding with Bartlett in 1858 in his capacity as editor of the Providence *Journal* asking him to publish "a song" for her. Mary Eastman to Bartlett, 7 January 1856, and 9 January 1858. JRB Papers, Correspondence.

15. Mary Eastman used the word "pecuniary" in letters to Henry Sibley, then a delegate to the U. S. Congress from Minnesota Territory, whom she pressed to remember "your promise to endeavor to get him ordered to Washington and if this is refused—perhaps you could succeed in getting him his furlough." Mary Eastman to Sibley, 20 December 1848, and 5 and 31 January 1849, Henry Hastings Sibley Papers, Minnesota Historical Society, St. Paul, reel 5:289-90, 349, 442 (HHS Papers).

16. That the Eastmans indicted Schoolcraft of duplicity is evidenced in his response to Eastman dated 28 April 1855. HRS Papers, reel 41:25263-64.

17. The inscriptions penciled on the verso of the watercolors consist of a descriptive title, indication of authorship "Drawn by Col. Eastman U.S.A. from sketch by J. R. Bartlett" (or a variation), and numbers assigned by The John Carter Brown Library (JCB) and by John McDermott in *Seth Eastman*. The exception is *Santa Cruz*, which bears its title within its narrow front margin. They appear to be in the same hand, not Eastman's (again, *Santa Cruz* is a possible exception). Eastman did not achieve the rank of Lieutenant Colonel until 1861 after he had left Washington, and McDermott's book was not published until 1961. These facts, however, do not explain the preciseness of the inscriptions, likely transcribed from another source. The margins (some trimmed) are unmarked, and there are no signs of erasure. What was found on the backs of the watercolors presumably informed RISD's cataloging, although the titles do not correspond exactly, and there are occasional misspellings; McDermott modified them slightly. Discoloration on the verso of the watercolors in broad bands on all four edges suggests that the watercolors were once pasted into a scrapbook.

18. Because the ordering of Eastman's watercolors assigned by the RISD Museum in 1947 and that used by The John Carter Brown Library and McDermott (*Seth Eastman*, 242-43), makes no sense in terms of Bartlett's boundary route, I have altered it completely. The sequence is now plausible.

19. *PN*, I:93.

20. So a Providence *Journal* article of 7 November 1850, "written by a gentleman attached to the main body of the Commission" pasted in Bartlett's Scrapbook, is entitled. On 1 November, they "traveled 13 miles to the Painted Caves, crossing the San Pedro or Devil's River twice,..." JRB Papers, Mexican Boundary Commission Scrapbook, 17, 18.

21. *PN*, I, illustration 98, discussion of crossing on 96-98. Bartlett's wash drawing is JCB64.

22. *PN*, I:181, II:568.

23. *PN*, I:232, 235. Bartlett described the Santa Rita on 384.

24. *PN*, I:232.

25. Bartlett, "Cañon, near Copper Mines, New Mexico," JCB85; Bartlett, "[Unidentified sketch of mountains, buildings, and river]," JCB131.

26. *PN*, I:227.

27. Bessau, "Copper Mines from the North," JCB178.

28. "La Bufa" means a volcanic extrusion. "Cobre" is given as "Cobra" in the inscription.

29. *PN*, I:235.

30. *PN*, I:416. Asa Gray, the botanist with the Commission, however, in his *PLANTAE NOVAE THURBERIANAE*....(1854), remarked on seeing, after leaving Bacuachi, a new cereus, "a species allied to C. giganteus, Engeln." JRB Papers, Mexican Boundary Commission Scrapbook, 301-02 (381-82).

31. Bartlett, "Cañon between Bacuachi and Arispe," JCB109. Eastman, "Cañon between [word missing] + Magdalena," Museum of Fine Arts, Boston, MA, 54.1690.

32. *PN*, I:278.

33. JRB Papers, Mexican Boundary Commission, Journal, 447.

34. Bartlett, "Magdalena, Sonora," JCB50. It informed the wood-cut with the same title of the town itself which Bartlett said he sketched on 3 October 1851; see *PN*, I:431.

35. Henry C. Pratt, *Bartlett Survey Traversing a Canyon*, 1855, Phoenix Art Museum, AZ.

36. *PN*, I:370, 364.

37. JRB Papers, Mexican Boundary Commission, Journal, 151-52.

38. "Piloncillo" is spelled "Pitoncillo" in the watercolor's inscription. To the title "Pass at the Pitoncillo [sic]" inscribed on the verso, "Chiricahui Mts in the Distance" was added in RISD's records (repeated by McDermott).

39. "Chiricahui Mtns.," JCB46; "Sept. 2, 1851. Pass through Sugar Mt. [Lf.] looking towards the Chiricahui Mts," JCB173j.

40. Bartlett, "La Punta de Sauz," JCB47.

41. *PN*, I:369, 370.

42. Whipple's report is Appendix D in *PN*, II:597-602.

43. Bartlett, "Great cañon of the Gila Oct. 27, 1851," JCB173a.

44. Whipple's report mentions early morning observations. While the instrument depicted in Eastman's watercolor appears to be a theodolite, a small telescope mounted so as to rotate on horizontal and vertical planes, which the Commission did carry with it, Robert V. Hine (*Bartlett's West: Drawing the Mexican Boundary* [New Haven and London: Yale University Press, 1989], 61) asserts that its use was impractical in inaccessible areas. For mapping techniques, see Willian H. Goetzmann, *Army Exploration in the American West*, Appendix B, 438-39, and Albert Boime, *Magisterial Gaze: Manifest Destiny and American Landscape Painting c. 1830-1865* (Washington, D.C. and London: Smithsonian Institution Press, 1991), 150-51 and Fig. 41, 152.

45. *PN*, I:401.

46. Ibid., opp. 402. Pratt, "Valley leading to Santa Cruz, Sonora," JCB4; Pratt, "Valley leading to Santa Cruz from the North," JCB26.

47. Ibid., 408. Bartlett's drawing is inscribed "for Woodcut. Santa Cruz - from the West, Sept. 25, 1851," JCB48.

48. Ibid., II:124-25.

49. A sketch of the fort nearby exists. Bartlett, "Fort Yuma - Junction of the Colorado + Gila rivers June 12, 1852," JCB172q.

50. "[Map of the Gila River from Rio San Pedro to Colorado River showing the route of Bartlett Expedition Nov. 5-24, 1852]," JCB164. The dates 5-24 November appear on the map itself but without the year; although 1852 has been assigned to it, the drawing must have been made by Whipple's party in 1851 as the locations with dates run east to west.

51. "Pima" appears in the watercolor's inscription.

52. Eastman to Schoolcraft, 19 October 1851, and 15 April 1852. HRS Papers, reel 40: 24193, 24318.

53. "Both tribes of Indians referred to, use a singular piece of frame-work made of three poles connected with a netting, which is carried on their backs." *PN*, II:236.

54. Eastman, "Pimo [sic] woman" and "Pima women weaving and Pimo [sic] huts," Peabody Museum of Anthropology and Ethnology, Harvard University, Cambridge, MA, 1 41-72 10-115 and 1 41-72 10 116. *PN*, II:225, 238.

55. *PN*, II:235. Pratt, "Maricopa Village," JCB29 (also inscribed "The Winter Village of the Pimo [sic] Indians the Rio Gila"). Bartlett admitted he could not distinguish between the Pimos [sic] and "their allies;" see *PN*, II:261.

56. *PN*, II:opp. 249. Bartlett, "Pimo [sic] villages + cultivated fields - with the desert between the Gila and the Salinas," JCB72.

57. Ibid., 369-70.

58. Bartlett, "Aug. 15, 1852. Mountains on Correlitas Road 40 miles from El Paso," JCB105.

59. Bartlett, "Ojo Caliente Oct 16 1852," JCB173h.

60. *PN*, II:406-07.

61. Bartlett, "From Luguna [sic] de Patos (looking west) Chihuahua the Laguna not in sight - Oct. 14, 1852," JCB173i.

62 The title is inscribed in this bi-lingual fashion on the back of the painting.

63. Eastman, "From the camp near Laguna del [sic] Patos Oct 14," Museum of Fine Arts, Boston, MA, 54.1685.

64. Bartlett, "[Unidentified mountains]," JCB147; and Bartlett, "Organ Mountains near Fort Fillmore September 27, 1852," JCB172j.

65. Even the list of "Subjects for illustrations" contain none, only curious topographical features, quotidian activities, and one accident. Among the first listed is "Missions near San Antonio." JRB Mexican Boundary Commission, Journal, 447.

66. Bartlett and Eastman went on to other pursuits. After publishing *Personal Narrative*, Bartlett returned to Rhode Island where he served as Secretary of State until 1873. As for Eastman, a commission for several paintings for the rooms of the House of Representatives Committee on Indian Affairs and the Committee on Military Affairs came his way in 1867 and 1870 respectively.

67. Turner to Bartlett, 31 November 1853. JRB Papers, Correspondence. Only a few months before, Bartlett had requested that his report "be published for the use of the Senate,

at as early a period as practicable" in a fashion corresponding to Schoolcraft's volumes. Hand-written statement dated in pencil 23 March 1853, JRB Papers, Mexican Boundary Commission Correspondence, 7:61.

68. Bartlett to Schoolcraft, 14 January 1850, Schoolcraft to A. S. Lowrey [?], 7 February 1850, and Bartlett to Schoolcraft, 29 December 1849. HRS Papers, reel 39:23639, 23659, 23602-03.

69. Henry Pratt responded in 1856 to Bartlett's "questions relating to your Boundary accounts" and five days later sympathetically added "The way in which you are treated in this matter of accounts is the most outrageous possible, and I wish you could postpone the settlement of them till a new administration when I am sure you would stand a better chance of decent treatment." In 1859 Bartlett received word that his claim for $288 incurred for "the rescue of Mexican captives from the Indians," during his service as Commissioner, was denied by the Treasury Department; Bartlett was certainly persistent. Pratt to Bartlett, 8 and 13 May 1856; the Treasury Department to Bartlett, 14 April 1859. JRB Papers, Correspondence.

70. Turner to Bartlett, 27 August 1857. JRB Papers, Correspondence.

71. Bartlett did have a streak of parsimony. Although he felt able to take on the second edition of his *Dictionary of Americanisms ...*, old debts remained. In a letter informing Bartlett that last proofs were being sent to him, Turner wrote (one more time) that as he had "never presented any bill for my work on the 'Personal Narrative,' I will now mention that my charge for revising and proof-reading is $150." The following month, Turner had to remind Bartlett again. Not until reminded by a spokesman for Turner's widow did Bartlett send payment. Another on the survey, Thomas Webb, also repeatedly asked for "back dues for Secretary's services, etc." Turner to Bartlett, 12 April and 20 May 1859, and a Mr. Randolph to Bartlett, 2 October 1860; Webb to Bartlett, 15 April, 6 July, 22 November 1861. JRB Papers, Correspondence.

72. *PN*, I: x. Schoolcraft requested "a quarto page" for the illustrations which "Capt Eastman will draw ... on stone" for the Indian history. Schoolcraft to Sibley, 1 February 1850, HHS Papers, reel 6:670.

73. *PN*, I: xii.

74. The Rhode Island School of Design, founded in 1877, opened the following year; Bartlett lived on Benefit Street next to present-day Benson Hall on the school's "campus," but the school was not established there until after Bartlett's death. I am grateful to Thomas R. Adams, Librarian emeritus, The John Carter Brown Library, for information on the location of Bartlett's Providence house.

SAM TRUETT

BORDERLANDS AND BORDER PLACES

The Historical Landscape of the U.S.-Mexico Border

In early January, 1853, John Bartlett and the remainder of his U.S.-Mexico Boundary Survey team sailed east from the coast of Texas on a steamer bound for New Orleans. These weary travelers had lived for over two years in a landscape visited by few Anglo Americans; a place of such striking curiosity and diversity that Bartlett later devoted almost six hundred pages to the experience in his *Personal Narrative of Explorations and Incidents in Texas, New Mexico, California, Sonora, and Chihuahua.* The account represented a first encounter with the borderlands for many of its readers. Through its words and images, a vast terrain between the Gulf of Mexico and the Pacific Ocean began to acquire detail in the mind's eye of an expanding nation. Valuable for its descriptions of border communities and their inhabitants, Bartlett's *Personal Narrative* also provided a detailed portrait of the lands these people called home. As such, the account remains indispensable for mapping the historical landscape of the region.

Bartlett's careful attention to the natural world had several motivations. On the most basic level, he was concerned with the survival of the survey team itself. Journal entries meticulously note the amount of water, firewood, and grass available to the men and wagon teams. Just as he relied on previous, often faulty, accounts to find these resources, so too did Bartlett sketch the lay of the land for future travelers and emigrants. He also read the landscape with an eye to the more distant future: judging the potential of the soil to sustain agriculture, assessing the amount of wood available to build railroads and homes, and most importantly, noting the presence or absence of water. At the same time, his portrayal drifted beyond these more practical considerations to capture the particular and peculiar. The details that he amassed on the natural history of the region—its myriad birds, snakes, lizards, prairie dogs, cacti, and even rock formations—not only reflected a 19th century passion for collecting and classifying, but also underscored for readers just how different the region was from the more "domesticated" eastern seaboard.

Above all else, Bartlett was struck by the land's aridity. Upon leaving the coastal plain of Texas, he wrote, "nature assumes a new aspect."[1] To the west, the soil dried out, water became harder to find, and the familiar greens quickly gave way to "that most unbecoming of all hues, yellow."[2] The animals of the borderlands, Bartlett noted, seemed quite at home in this strange land: "it will have been observed how beautifully nature has adapted them for these districts." He was more apprehensive about the land's potential for humans. "Here man ... cannot live;" he concluded, "for there is no soil that he can cultivate, no water to slake his thirst, no wood to supply him with fuel; nor can the domestic animals so necessary to him exist."[3] Bartlett doubted that the region would ever support communities that his countrymen could call home. "Is this the land which we have purchased, and are to survey and keep at such a cost?" he asked himself at one point. "As far as the eye can reach stretches one unbroken waste, barren, wild, and worthless."[4]

Portrayals of the natural world as lists of resources, catalogs of species, or simply "barren waste" were typical of Anglo American accounts of the region. Although they shed light on the mindset of 19th century American exploration, they are less useful for understanding the land itself because they are incomplete: they describe the borderlands as if human residents did not matter. A close reading of Bartlett's account, though, shows that he traversed a profoundly human place. Rarely did a day pass that he did not encounter the world of people—people who had lived on the land for years and had reshaped it into a mosaic of hunting grounds, roads, fields, fences, mines, homes, and even cities. To understand the environmental history of the U.S.-Mexico borderlands, one must follow the story of those who lived there. For as Bartlett himself noted, the land that had been taken from Mexico was a land haunted by the ghosts of those who had made nature their own for millennia.

Ghost Landscapes

The first human visitors to the borderlands were the precursors of present-day American Indians. Centuries ago, the region was much wetter and cooler than it was when Bartlett surveyed it. Piñon-juniper forests and lush grasslands blanketed much of the arid landscapes depicted in the survey paintings of Henry Cheever Pratt. Many plants associated with the contemporary American Southwest—the saguaro and the foothills palo verde—had yet to migrate north from Mexico.[5] One can only guess what the first peoples to encounter this land thought about it; perhaps they found it as strange and wonderful as Bartlett did. Fragmentary remains of their story suggest that, whatever their initial opinions, they found the place hospitable enough to put down roots. Like all landscapes, this one soon bore the distinctive signature of those who called it home.

A clue to this early human landscape emerged in the 20th century, when desert rains uncovered the 11,000-year-old skeleton of a woolly mammoth near the border town of Naco, Arizona. Among its bones, archaeologists found several stone spear points, evidence that the animal had probably been killed by human hunters.[6] This page of history, etched in dry desert clay, sheds light on early human lifeways, and on possibly the first major episode of man-made environmental destruction in the borderlands. Shortly after the Naco kill, an entire menagerie of large mammals, including mammoths, camels, horses, and giant sloths, vanished from the region without a trace. Although some argue that climate changes after the last ice age were responsible for these extinctions, others point to the ghostly remains at Naco and similar sites. Here, they contend, long before market hunters swept the bison from the Great Plains, the first American settlers brought about a similar disaster. While the mystery remains unsolved, the mounting evidence suggests that people and climate may have collaborated in these deaths.[7]

These extinctions, along with post-ice age climate warming, brought critical changes to the human landscape. Where people once followed the footprints of large mammals, new immigrants to the region traced a broader range of pathways, moving seasonally between desert lowlands, mountain forests, and ever-shrinking bodies of water in the search for food. In this way, they slowly developed skills to survive in the arid lands that the earlier hunters had circumvented, and that later travelers like Bartlett would consider uninhabitable.[8] Although these newcomers had to keep in step behind nature's cycles, they were far from slaves to a harsh environment. Some stored surpluses to carry them through lean times. Others fashioned snares and traps to supplement their hunting and gathering efforts and even sought to change biology itself by transforming wolves into domestic hunting dogs.[9] Whatever the "natural" limitations to human survival, human-land relations were ultimately rooted in cultural strategies, often aimed at increasing the food supply and reducing the work and space needed to obtain it.

About three thousand years ago, border peoples made a radical breakthrough in the way they approached this economic puzzle, with environmental consequences as profound as those of the post-ice age extinctions. At the center of these changes was corn, a product of genetic manipulation that appeared in central Mexico about five thousand years ago, and that later moved north along native trade routes. As corn came to supplement seasonal hunting and gathering, border peoples shifted from a life of food gathering to one that also relied on food production. Places that had previously been lived in seasonally were gradually remade into more "settled" landscapes of permanent homes, irrigation canals, fields, roads, and trade centers.[10] One such landscape developed along the Gila and Salt Rivers of Arizona around fifteen hundred years ago, when a group known today as the Hohokam began building a network of villages and canals that extended hundreds of miles into the surrounding desert. The level of organization needed to maintain this water control system—the most extensive in North America at the time—could only be attained by seriously shifting relations of human power. At its height, Hohokam society became highly stratified and authoritarian, bolstering interwoven systems of environmental and social control that would only be surpassed in the 20th century with the rise of federal hydraulic power projects and their attendant bureaucracies.[11]

Bartlett was struck by the complexity he saw in the vestiges of the Hohokam world: "The ruined buildings, the irrigating canals, and the vast quantities of pottery of a superior quality, show that, while they were an agricultural people, they were much in advance of the present semi-civilized tribes of the Gila."[12] One can appreciate Bartlett's sense of wonder about these ghost landscapes. Multi-storied Hohokam ruins, bearing for Bartlett "a striking resemblance to ... the site of ancient Babylon," resonated more with European notions of civilization and progress than did the wattled fences and outdoor ramadas of the Pima and Maricopa Indians.[13] What Bartlett considered most "advanced" about the Hohokam might have been critical to this culture's demise. Although they continued to rely on hunting and gathering, especially in the years when droughts and floods made their irrigation systems useless, the Hohokam people became increasingly committed to their hydraulic regime. A greater dependence on a narrower range of resources, as well as the increased centralization required to maintain their irrigation economy, made the Hohokam more vulnerable to social and environmental disruptions. Some scholars argue that several hundred years before the Spanish arrived, increased social strife, perhaps combined with drought, disrupted the delicate balance between these desert people and their resources. By 1300 A.D., they had abandoned their agricultural oases. Maybe the ghost landscapes they left behind served as ecological object lessons for others; never again did native economies approach the scale of the Hohokam.[14]

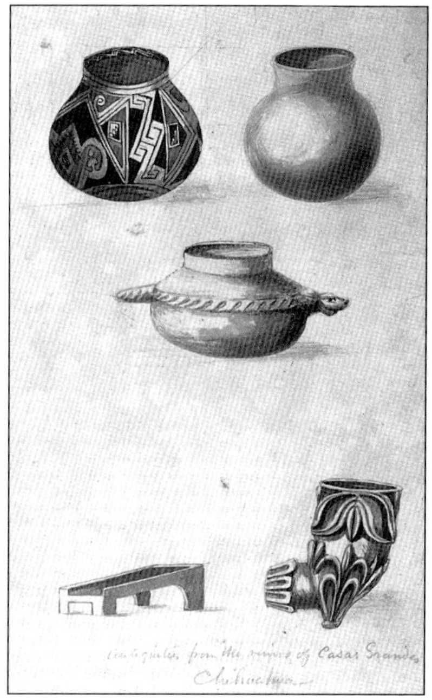

Figure 44
John Russell Bartlett
Antiquities from the Ruins of Casas Grandes, Chihuahua, 1852
9" x 5 7/8" • Pencil and ink
The John Carter Brown Library, Brown University, Providence, Rhode Island [79]

By the time the first Europeans arrived on the scene, native peoples had developed an amazing variety of strategies for living in the coastal plains, semi-arid hills, arid plateaus, grasslands, and deserts that stretched between the Gulf of Mexico and the Pacific Ocean.[15] The first written account of the land and its people was that of Alvar Núñez Cabeza de Vaca, a Spaniard who was shipwrecked off the coast of Texas in 1528. For eight years Cabeza de Vaca traveled west amongst the region's native peoples, walking six thousand miles across much of the same terrain later visited by Bartlett. By the time he was "discovered" by Spanish slave traders in present-day Sinaloa, he had learned much about the area's climate, vegetation, animals, and cultures.[16] The 1542 account of his adventures provided the Spanish with their first glimpse of the vast "terra incognita" to the north, and stimulated later expeditions into the region in search of treasure, land, and souls.[17]

Reading Cabeza de Vaca, one is struck by just how *human* this terra incognita was. The lagoons, coastal prairies, and forested valleys he visited in east Texas were noted for their diversity of plant and animal life, precisely because this diversity kept the Spaniard and his native cohorts alive at the different seasons of the year. Farther to the west, nature's capacity to sustain human life greatly diminished. Moving across arid plateaus and deserts, Cabeza de Vaca learned from the Indians to read the land for signs of edible life. He learned the value of the prickly pear fruit, and farther west, of pine-nut mash and mesquite-bean flour. Much of his journey was a struggle to stay alive in a strange land where the line between feast and famine was thin. "Occasionally, these Indians kill deer and take fish," he wrote of one group, "but the quantity is so small and famine so prevalent that they eat spiders and ant eggs, worms, lizards, salamanders, snakes,

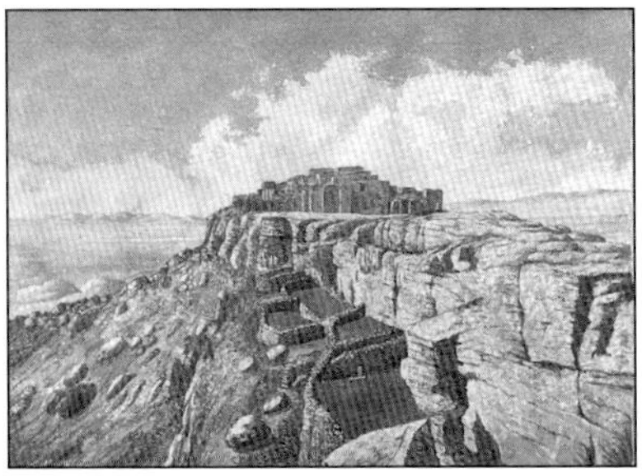

Figure 45
Henry Cheever Pratt
Moqui Village, Arizona, undated
10" x 14" • Watercolor
The John Carter Brown Library, Brown University, Providence, Rhode Island [20]

and poisonous vipers; also earth and wood—anything, including deer dung and other matter I omit."[18] Although less hospitable than the coastal plains of Texas, these western lands were clearly more than barren waste. As Cabeza de Vaca learned with his stomach, these were places that, at one season or another, were remade into the most essential of human landscapes.

The native transformation of the land was perhaps more obvious to explorers who marched north into what is today New Mexico. In 1540, when Francisco Vásquez de Coronado visited the Pueblo villages of the upper Rio Grande Valley, he found a land that differed notably from that which sustained the more nomadic peoples to the south and east. Here the land clearly bore the mark of the human labor that transformed it. The Pueblo Indians had planted the desert with corn and cotton, and lived in permanent villages—much as the Hohokam of Arizona had done before.[19] Like their neighbors, the Pueblos influenced a vast territory not immediately recognized by the Spanish as human. As they maintained their farms along the Rio Grande, they also wandered far afield to hunt deer and bison, to collect plants, and to trade with Plains Indians for meat and hides they could not obtain themselves.[20]

At the time of European contact, a large portion of the native borderlands fell into either of two patterns. Where a year-round supply of water permitted it, people could seek a more settled relationship to the land, as the Pueblo Indians did. The Maricopas and Pimas of the Gila River and the Yuman-speaking groups of the Colorado River, both central to Bartlett's account, fell into this category. More typically, however, people followed the seasons across a wide range of ecological communities, much as Cabeza de Vaca had done in Texas. This was true for groups like the Pueblos and Pimas, but it was essential for desert and mountain-living peoples like the Papagos and Apaches.[21] Distinct in their local adaptations to the land, these various peoples mapped their relationship to the natural world according to their particular habits, beliefs, and histories. In all these ways they made a place their own. At the same time, networks of trade, warfare, marriage, and religion set linkages between desert plain, river valley, and mountain highland, blurring the borders between one native landscape and the next.

A similar mosaic of connection and division would mark the new borderlands between European and Indian people—borderlands that emerged in the 16th and 17th centuries, when New Spain expanded north into the present-day states of Sonora, Chihuahua, Arizona, New

Mexico, and California. Like the people already there, the Spanish had their own special strategies for reshaping nature into a human home—strategies that had been forged in the arid and semi-arid reaches of Spain, Northern Africa, and highland Mexico. Yet unlike the native Americans, who based their lifeways around centuries of adaptation to the region's ecological communities, the Europeans brought with them the plants and animals of their Old World landscapes.[22] The consequences of this seemingly trivial fact were revolutionary. The subsequent history of the region can hardly be told without reference to these Old World species.

Ecological Conquests

The most significant of these non-human migrants were the various germs responsible for diseases such as smallpox, rubella, and malaria, microscopic species that followed Europeans, Africans, and their animals west across the ocean. Often spreading to the northern frontiers of Spanish Mexico before the colonists themselves arrived, these germs devastated New World peoples, who carried no natural antibodies to their associated diseases. An epidemic could destroy up to 90% of a given community, shattering the political and economic networks that linked it to the land. Death and disorder, in turn, made native groups more vulnerable than ever to cultural and territorial conquests.[23] The first documented epidemics in the border region followed the slave-raiding expeditions of Nuño de Guzmán in the early 1530s—the same expeditions, in fact, that met with Cabeza de Vaca at the end of his journey. By the time the first written accounts of the region were being generated, depopulation by disease was already a fact of life in northern New Spain. The lands described by Spanish travelers were ghost landscapes, not only by virtue of the agonizing deaths that haunted the place, but also because the land itself still bore the physical mark of the hundreds of communities that had suddenly vanished, as if into thin air.[24]

As epidemics depopulated communities and disrupted connections to the land, other European organisms created opportunities for native Americans. By trading with Spanish settlers, or by resettling onto mission lands, Indian groups like the Lower Pimas of Sonora and the Gila River Pimas of Arizona began sowing domesticated plants from Europe and central Mexico. Native crops of corn, squash, beans, and cotton were, in this fashion, supplemented by such exotic cultivars as wheat, fruit trees, watermelons, tomatoes, and chiles.[25] The benefits of new species were felt by both Indians and Europeans. Old World species increased crop diversity, lengthened growing seasons, and often increased the scope of native economies. The Pima Indians of the Gila River, for instance, found that growing wheat for themselves and for Hispanic markets allowed them to gain population and political power.[26] Spanish settlers, for their part, discovered that native beans, corn, and squashes added to the productivity and stability of their fields. Particularly in areas distant from Spanish markets, the cultural exchange of crops, technology, and farming strategies, often a counterpart to racial mixing, was a key to frontier survival.[27]

The introduction of Old World animals contributed even more dramatically to this process of social change. Although cattle and horses were key participants of the Spanish *entrada* into the region, both species migrated north ahead of the settlers, in some areas becoming part of the native landscape long before Europeans gained control of the land. The Comanche Indians began using the horse for hunting, transportation, warfare, and even food—a move that helped them resist the territorial conquests suffered by their Indian neighbors. The Papagos and Apaches, on the other hand, immediately added Spanish cattle to the list of animals that could be hunted when meat was needed.[28] Like Old World plants, these animals often occupied a "hybrid" landscape where European and Indian worlds met, struggled, and sometimes merged. In the contested frontier between New Spain and the Apachería, for instance, European and Mestizo settlers adopted the custom of hunting wild cattle in the manner of their Indian enemies—partly because Apache raids made herding impracticable. The wild cattle and Mexican cattle hunters that Bartlett later met along Arizona's San Pedro River were testimony to this enduring "borderlands" landscape—a place as much native as European.[29]

Key agents of economic and cultural change, Old World animals also contributed to the transformation of the land itself. Not since the ice-age extinctions had so many large herbivores permeated the region's arid grassland and riverine ecosystems. As cattle, sheep, and horses moved across the land, their hooves cut into the fragile soil, creating well-worn trails that quickly became gullies after a season of desert rains. In the most heavily grazed areas, gullying eroded fertile topsoil, stimulating the replacement of native grasses with low-quality grasses, native shrubs, and the exotic weeds that had often evolved alongside cattle in Old World fields and pastures. In broader perspective, however, these ecological impacts remained small in comparison to those of American cattle ranching in the region in the years after Bartlett's survey.[30] Even though Spanish herds in the Moctezuma and Sonora river valleys of Sonora numbered in the thousands by the late 17th century, instability of regional markets, combined with Apache raiding, may have prevented them from staying in any one place long enough to permanently disrupt desert ecosystems.[31]

By far the greatest ecological impacts of Old World plant and animal species were their contributions to human population growth. A broader array of crops increased the agricultural productivity of the river bottoms, while cattle and sheep expanded the range of plant communities that contributed—albeit indirectly—to the human economy of the borderlands. Equally critical for transforming the desert into a populated landscape of towns, pastures, and fields, mines, and smelters, were commercial linkages between the northern frontier and the central Mexican plateau. By the late 17th century, the rise of such connections around one resource, silver, would have momentous consequences for regional development. As Indian and Mexican workers migrated to the mining centers of San Juan Bautista and Nacozari, Sonora, markets for crops and livestock from nearby river valleys grew.[32] This demand coincided with a growing secularization of land and development of roads to transport resources through emerging market centers like Guaymas, Sonora and Parral, Durango to Mexico City. In a few short decades, ranchers, farmers, merchants, and miners began to appropriate Indian and mission lands all across northern Mexico, initiating a piecemeal reorientation of natural resources and human labor more fully into colonial capitalist markets.[33]

By the early 19th century, these capitalist linkages began to tug the lands and peoples of the region in a new direction. Following its independence from Spain in 1821, Mexico removed colonial trade restrictions, making Santa Fe a port of entry for goods from the United States. For the next several decades, American wagon caravans rattled west across the plains, bringing with them manufactured goods from the eastern United States and Europe, and returning home with Mexican silver and mules.[34] Following these paths west, Anglo Americans discovered other elements of the borderlands environment that could be transformed into economic wealth. In the mid-1820s, American fur trappers fanned across New Mexico, and out along the Gila River of Arizona, in search of beavers that made their home in the river valleys. Within a decade, this mammal was all but eradicated from the region.[35] As beaver dams slowly crumbled and vanished from the landscape, river oases became more vulnerable than ever to flooding, channel cutting, and desertification.[36] Slowly but surely, new markets, whether in animals or in minerals, were setting environmental change to the beat of a new drummer from the north.

In other parts of the borderlands, however, these new bonds would have to wait until after the American conquest of Northern Mexico to fully materialize. In Sonora and eastern Chihuahua, economic and political instability forced the abandonment of numerous mines and ranches by the mid-18th century. Conditions worsened with the renewed incursions of Apache Indians. For this nomadic tribe, the rise in Spanish livestock numbers, both on ranches and on trans-sierran ore trains, represented new subsistence opportunities; killing or stealing mules and cattle was often easier than hunting wild game. The Spanish responded by reorganizing border defenses in the military colonies, or presidios, of Janos, Fronteras, and Tucson. For several decades, Spanish soldiers held their ground, sometimes by fighting, but just as often by "pacifying" the Apaches with bequests of horses and food, and invitations to settle on lands nearby.[37] The flow of money for soldiers and gifts shrank to a mere trickle by the early 19th century, as the Mexican wars of independence depleted the national treasury. By the 1830s, renewed Indian raids deep into the northern frontier began to unsettle the countryside. One by one, ranching, agriculture, and mining settlements were abandoned, leaving in their passage the mere ghosts of landscapes past.[38]

Looking Backward

Bartlett's *Personal Narrative* is haunted by these ghosts at almost every turn. Just as early Spanish colonists entered a land that had been widowed by Old World disease, Bartlett's path led through the remains of prior settlement: crumbling walls and bullet-scarred missions attesting to the depredations of the Apaches, and of the new "barbarians" from the north, the Texans and Americans. These ruins evoked mixed emotions of wonder and despair. Upon entering Fronteras, Sonora, he wrote: "the church forms the prominent object in the landscape, and its style is quite picturesque; its effect also is heightened by its somewhat ruined condition." On closer inspection, however, he changed his mind, adding that "one's ideas of the picturesque are soon dissipated by the sight of its ruined adobe buildings."[39] This sense of loss was echoed in the countryside, where haciendas and ranchos seemed to have returned to a state of wildness. At the San Bernardino hacienda on the present-day Arizona border, Bartlett noted that cattle had "greatly multiplied since, and now roam over the plains and in the valleys, as wild and more fierce than the buffalo."[40] Farther south in Sonora, the estate of nature unfolded before him, as he found bats and swallows in "full possession" of church edifices; ranchos and corrals "overgrown with tall grass and shrubbery"; and abandoned fruit trees "bending with their loads of fruit." His romantic delight in nature notwithstanding, Bartlett concluded that "it was really sad to see so beautiful a region reverting to the condition of a wilderness."[41]

What did these ghost landscapes signify to Bartlett? Why spend so much ink on an exploration of new national territory, dwelling on the symbols of another nation's decay? A clue to this riddle can be found in the words of another American, Thomas Robinson Warren, who lived in Sonora at the time. "It was enough to make one's blood boil," Warren wrote in 1858, "to see those beautiful plains a vast solitude—extensive ranchos abandoned and uninhabited, and villages in ruins." All across this "natural paradise," he continued, "one may not meet a human being, although on every side he may see fine habitations lying waste."[42] One senses a backward-looking romanticism in Warren's voice, a nostalgia for the era before Spain's fall from this "natural paradise." Yet reading on, one realizes that this abandoned landscape is actually the backdrop for a different, more progressive story: the prophecy of America's "reclamation" of the region. Warren elaborates on Sonora's farmlands and mines, "utterly devastated and given over to wandering tribes of savages," and then moves to his broader conclusion: "Does not the certainty suggest itself, that before many years the American people, proverbially ambitious and restless as they are, will spread themselves over this hidden paradise?"[43]

From this perspective, the borderlands emerge as a space between nations and as a set of linkages spanning the distance between past and future. This, in the end, is what the readers of Bartlett would have understood. The lists of resources, the exotic plants and animals, even the barren wastes: these were all part of a new American promise of opportunity that had its roots not in virgin wilderness, but rather in the soil of failed landscapes. Thus begin two distinct stories about a common continental landscape, mythical tales of cultural difference that work even today to divide and distort the overlapping histories of the U.S.-Mexico borderlands. One would do well to move beyond these illusions to revisit the ghosts that live on in the land and its people. Though surveyed by Bartlett well over a century ago, the region remains terra incognita, a broad expanse of opportunity for those willing to make the two thousand mile-long journey, looking backward.

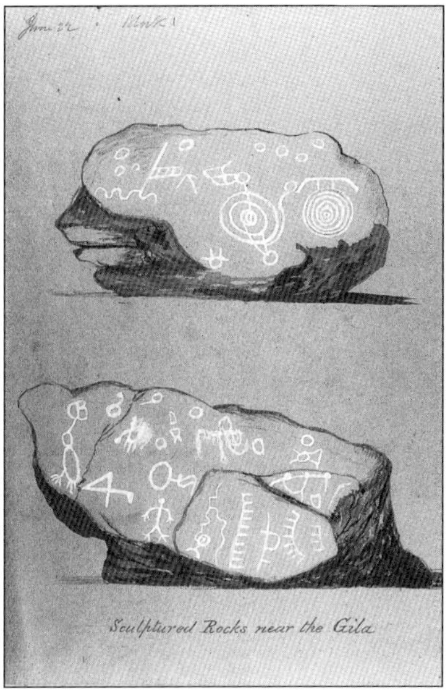

Figure 46
John Russell Bartlett
Sculptured Rocks near the Gila, 1852
8 1/2" x 11 1/2" • Sepia wash
The John Carter Brown Library, Brown University, Providence, Rhode Island [154]

NOTES
Borderlands and Borderplaces: The Historical Landscape of the U.S.-Mexico Border

1. John Russell Bartlett, *Personal Narrative of Explorations and Incidents in Texas, New Mexico, California, Sonora, and Chihuahua*, 2 vols. (New York: D. Appleton, 1854), I:138.

2. Bartlett, *Personal Narrative*, I:248.

3. Bartlett, *Personal Narrative*, II:562-63.

4. Bartlett, *Personal Narrative*, I:247.

5. Julio L. Betancourt, Thomas R. Van Devender, and Paul S. Martin, eds., *Packrat Middens: The Last 40,000 Years of Biotic Change* (Tucson: University of Arizona Press, 1990), 2-10; 134-62. My treatment of this early history parallels that of Thomas E. Sheridan, *Arizona: A History* (Tucson: University of Arizona Press, 1995).

6. See Emil W. Haury, "Artifacts with Mammoth Remains, Naco, Arizona: Discovery of the Naco Mammoth and the Associated Projectile Points" and related essays in J. Jefferson Reid and David E. Doyel, eds., *Emil W. Haury's Prehistory of the American Southwest* (Tucson: University of Arizona Press, 1986).

7. For the best discussion of the human and climatic factors behind these extinctions, see the essays in Paul Martin and Richard Klein, *Quaternary Extinctions: A Prehistoric Revolution* (Tucson: University of Arizona Press, 1984).

8. Robert H. Lister and Florence C. Lister, *Those Who Came Before: Southwestern Archaeology in the National Park System* (Tucson: University of Arizona Press, 1984), 16-18.

9. Arrell Morgan Gibson, *The American Indian: Prehistory to the Present* (Lexington, MA: D.C. Heath and Company, 1980), 22-24.

10. Lister and Lister, *Those Who Came Before*, 18-52. See also Linda S. Cordell, *Prehistory of the Southwest* (New York: Academic Press, 1984), 181-243.

11. See essays in George J. Gumerman, ed., *Exploring the Hohokam: Prehistoric Desert Peoples of the American Southwest* (Albuquerque: University of New Mexico Press, 1991) and Patricia L. Crown and W. James Judge, eds., *Chaco & Hohokam: Prehistoric Regional Systems in the American Southwest* (Santa Fe: School of American Research Press, 1991). For issues of social and environmental control in 20th century water projects, see Donald Worster, *Rivers of Empire: Water, Aridity and the Growth of the American West* (New York: Pantheon Books, 1985).

12. Bartlett, *Personal Narrative*, II:277.

13. Bartlett, *Personal Narrative*, II:245.

14. Crown and Judge, *Chaco and Hohokam*, 296-308; Sheridan, *Arizona*, 16-19.

15. The best source on border ecological communities west of El Paso is David E. Brown, ed., "Biotic Communities of the Southwest: United States and Mexico," *Desert Plants* 4:1 (1982): 1-341. See also Frederick R. Gehlbach, *Mountain Islands and Desert Seas: A Natural History of the U.S.-Mexican Borderlands* (College Station: Texas A&M University Press, 1981).

16. Alvar Núñez Cabeza de Vaca, *Cabeza de Vaca's Adventures in the Unknown Interior of America*. Trans. and ed. by Cyclone Covey (New York: Crowell-Collier, 1961; repr., Albuquerque: University of New Mexico Press, 1983).

17. Cabeza de Vaca, *Cabeza de Vaca's Adventures*, 12-13.

18. Cabeza de Vaca, *Cabeza de Vaca's Adventures*, 79.

19. For Pueblo culture and early Spanish contact, see Edward H. Spicer, *Cycles of Conquest: The Impact of Spain, Mexico, and the United States on the Indians of the Southwest, 1533-1960* (Tucson: University of Arizona Press, 1962) 152-58. For the Coronado expedition, see George P. Hammond and Agapito Rey, eds. and trans., *Narratives of the Coronado Expedition, 1540-1542* (Albuquerque: University of New Mexico Press, 1940).

20. Katherine A. Spielmann, "Colonists, Hunters, and Farmers: Plains-Pueblo Interaction in the Seventeenth Century," in David Hurst Thomas, ed., *Columbian Consequences*, Vol. I, *Archaeological and Historical Perspectives on the Spanish Borderlands West* (Washington: Smithsonian Institution Press, 1989), 101-13.

21. This is a simplification of social patterns outlined in Spicer, *Cycles of Conquest*, 8-15.

22. For an overview of these processes, see David J. Weber, *The Spanish Frontier in North America* (New Haven: Yale University Press, 1992), 302-34.

23. Daniel T. Reff, *Disease, Depopulation, and Culture Change in Northwestern New Spain, 1518-1764* (Salt Lake City: University of Utah Press, 1991). Also see general discussion of Old World diseases in Alfred W. Crosby, *Ecological Imperialism: The Biological Expansion of Europe, 900-1900* (Cambridge: Cambridge University Press, 1986).

24. Reff, *Disease*. See also Henry F. Dobyns, *From Fire to Flood: Historic Human Destruction of Sonoran Desert Riverine Oases* (Socorro, NM: Ballena Press, 1981), 45-56.

25. Weber, *Spanish Frontier*, 303-09; Robert C. West, *Sonora: Its Geographical Personality* (Austin: University of Texas Press, 1993), 27-43; and Paul H. Ezell, "The Hispanic Acculturation of the Gila River Pimas," *American Anthropologist* 63 (October, 1961):1-171.

26. Ezell, "Hispanic Acculturation," 134-35.

27. Weber, *Spanish Frontier*, 326-34; Ignaz Pfefferkorn, *Sonora: A Description of the Province*. Trans and annot. by Theodore E. Treutlein (Tucson: University of Arizona Press, 1989).

28. Weber, *Spanish Frontier*, 304-310; Richard White and William Cronon,"Ecological Change and Indian-White Relations," in Wilcomb E. Washburn, ed., *History of Indian-White Relations*, Vol. 4, *Handbook of North American Indians* (Washington: Smithsonian Institution, 1988), 417-29; Robert A. Hackenberg, "Pima and Papago Ecological Adaptations," in Alfonso Ortiz, ed., *Southwest*, Vol. 10, *Handbook of North American Indians* (Washington: Smithsonian Institution, 1983), 161-77; and Paul Hirt, "The Transformation of a Landscape: Culture and Ecology in Southeastern Arizona," *Environmental Review* 13:3/4 (Fall/Winter, 1989): 167-89.

29. Bartlett, *Personal Narrative*, I:398.

30. Weber, *Spanish Frontier*, 310-11; Dobyns, *Fire to Flood*, 57-89; Conrad Joseph Bahre, *A Legacy of Change: Historic Human Impact on Vegetation of the Arizona Borderlands* (Tucson: University of Arizona Press, 1991), 109-23; and Ronald U. Cooke and Richard W. Reeves, *Arroyos and Environmental Change in the American South-West* (Oxford: Clarendon Press, 1977), 24-99.

31. West, *Sonora*, 58-59, 70-78; Bahre, *Legacy*, 114-15.

32. West, *Sonora*, 44-66.

33. West, *Sonora*, 67-69; Robert C. West, "The Mining Community in Northern New Spain: The Parral Mining District," *Ibero-Americana* 30 (1949); Stuart F. Voss, *On the Periphery of Nineteenth-Century Mexico: Sonora and Sinaloa, 1810-1877* (Tucson: University of Arizona Press, 1982).

34. David J. Weber, *The Mexican Frontier, 1821-1846; The American Southwest Under Mexico* (Albuquerque: University of New Mexico Press, 1982), 122-29.

35. Weber, *The Mexican Frontier*, 131-33. See also David J. Weber, *The Taos Trappers: The Fur Trade in the Far Southwest, 1540-1846* (Norman: University of Oklahoma Press, 1971).

36. See Amadeo M. Rea, *Once a River: Bird Life and Habitat Changes on the Middle Gila* (Tucson: University of Arizona Press, 1983).

37. Rea, *Once a River*, 70-78. See also Weber, *Mexican Frontier*.

38. Weber, *Mexican Frontier*, 83-121.

39. Bartlett, *Personal Narrative*, I:265.

40. Bartlett, *Personal Narrative*, I:256.

41. Bartlett, *Personal Narrative*, I:414-15.

42. Thomas Robinson Warren, *Dust and Foam, or Three Oceans and Two Continents* (New York: Charles Scribner Press, 1858), 202.

43. Warren, *Dust and Foam*, 216-17.

COLOR PLATES

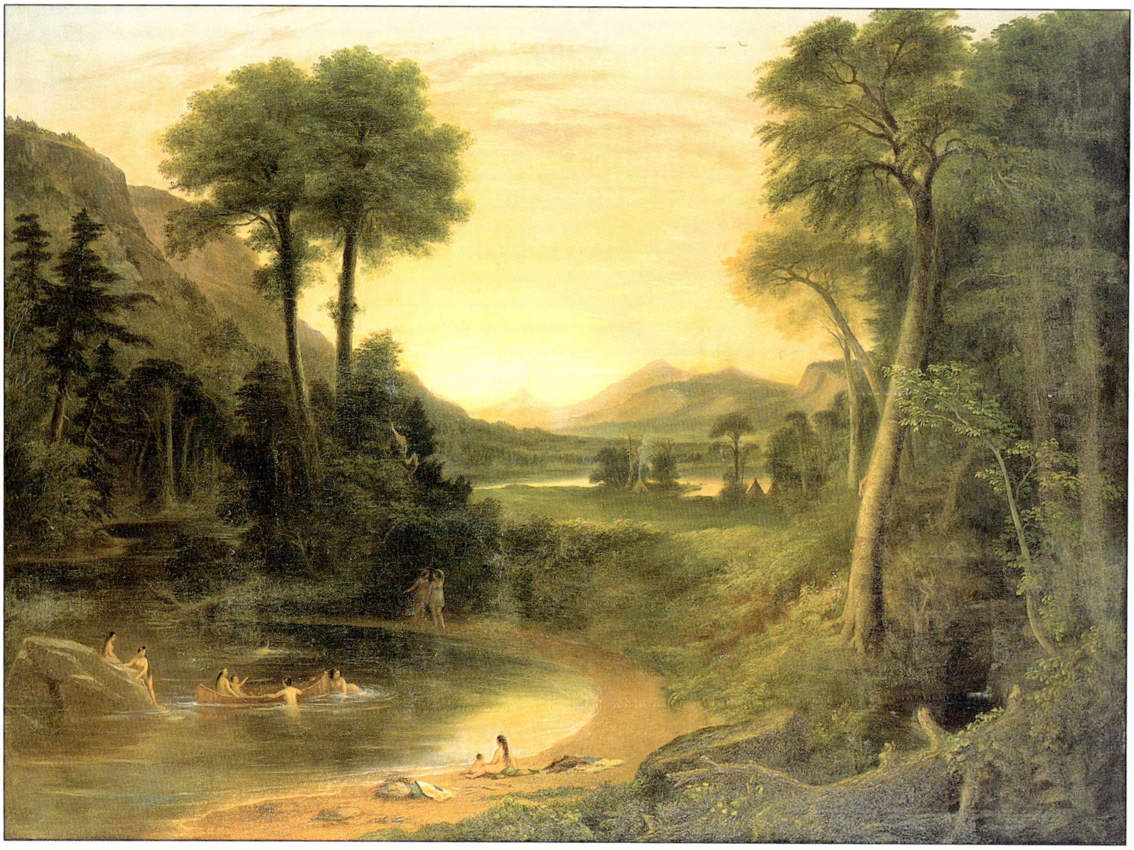

Plate 1
Henry Cheever Pratt
Indians Bathing: A Scene of New England, 1843
27 1/2" x 37 1/2" • Oil on canvas
Loan courtesy of William Blundin, New York, New York
Photo courtesy of Cooley Gallery, Old Lyme, Connecticut

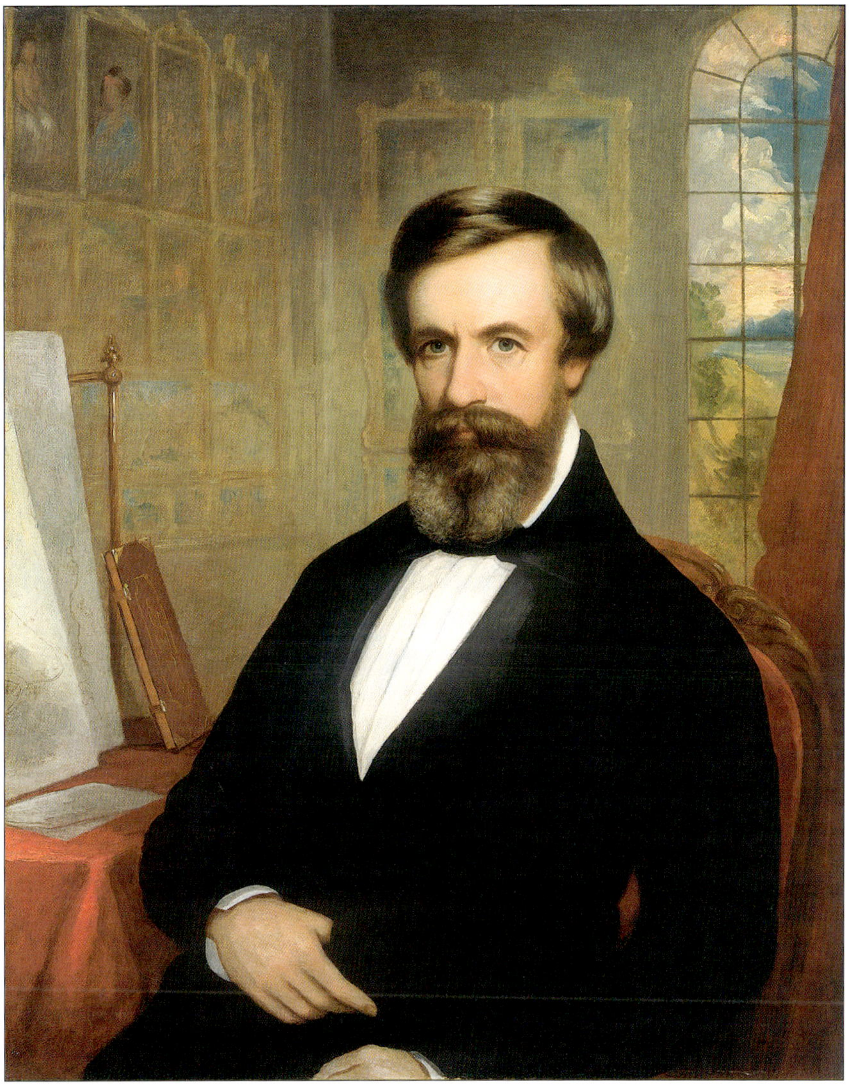

Plate 2
Henry Cheever Pratt
Portrait of John Russell Bartlett,
October 1, 1852
36 1/4" x 29 1/4" • Oil on canvas
Loan courtesy of Amon Carter Museum, Ft. Worth, Texas
Photo: L. Lorenz

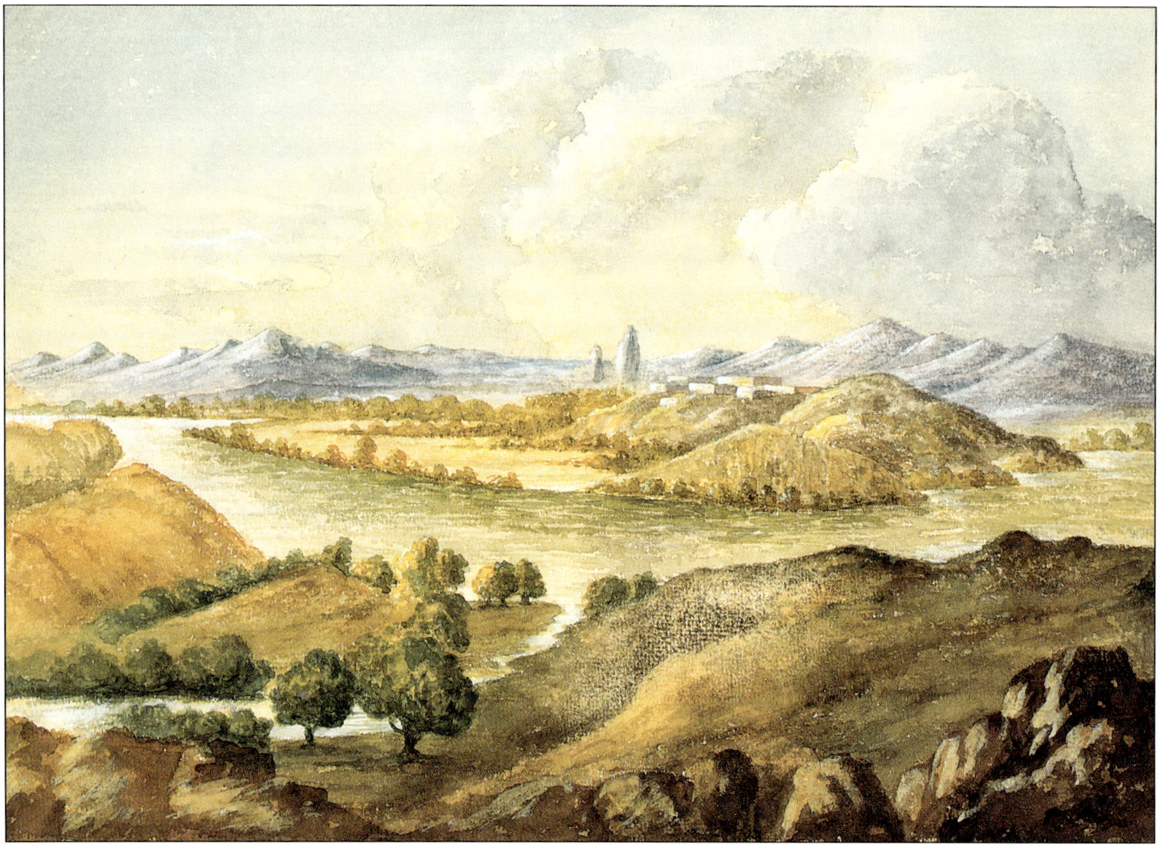

Plate 3
Henry Cheever Pratt
Fort Yuma, Colorado River, June 1852
9 3/4" x 9 1/8" • Watercolor
The John Carter Brown Library, Brown University, Providence, Rhode Island [1]
Photo: Damian Andrus

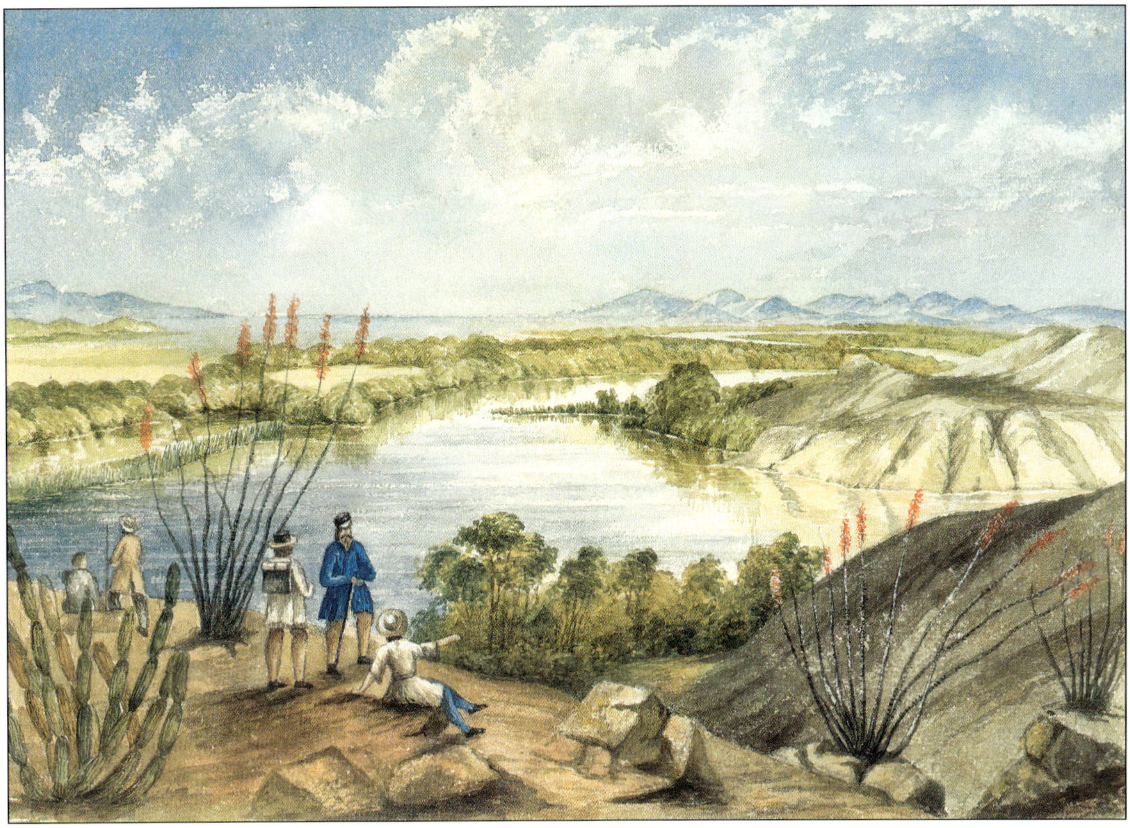

Plate 4

Henry Cheever Pratt

Junction of the Colorado and the Gila Rivers, Looking up the Gila

June 16, 1852

9 3/4" x 14" • Watercolor

The John Carter Brown Library, Brown University, Providence, Rhode Island [7]

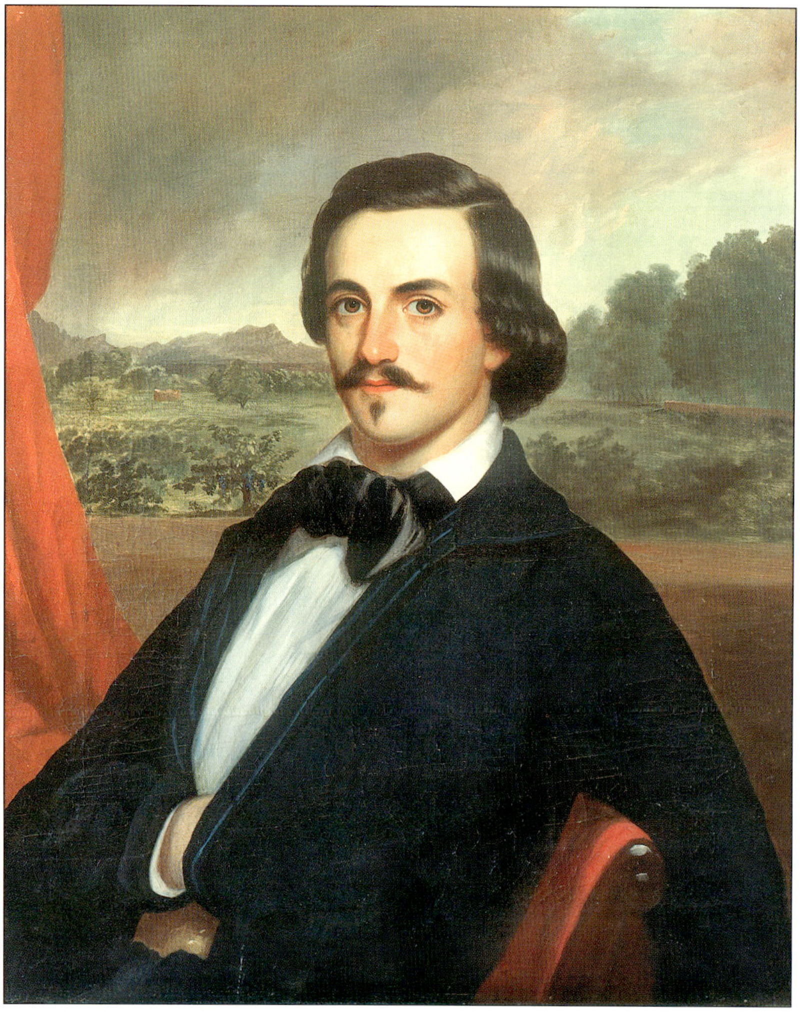

Plate 5

Henry Cheever Pratt

Portrait of James Wiley Magoffin,
October 1852

36" x 28 3/4" • Oil on canvas

Loan courtesy of Magoffin State Historic Site, El Paso, Texas

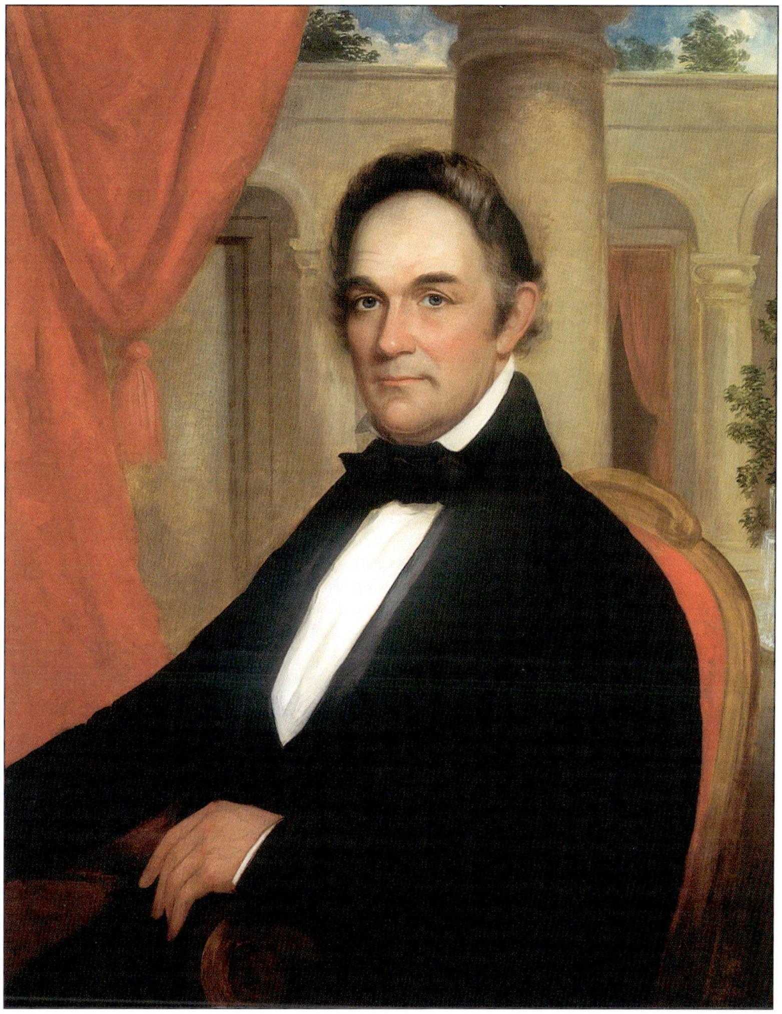

Plate 6
Henry Cheever Pratt
Portrait of Benjamin Franklin Coons, 1851
30" x 37" • Oil on canvas
Loan courtesy of Philip and Amy Alfeld

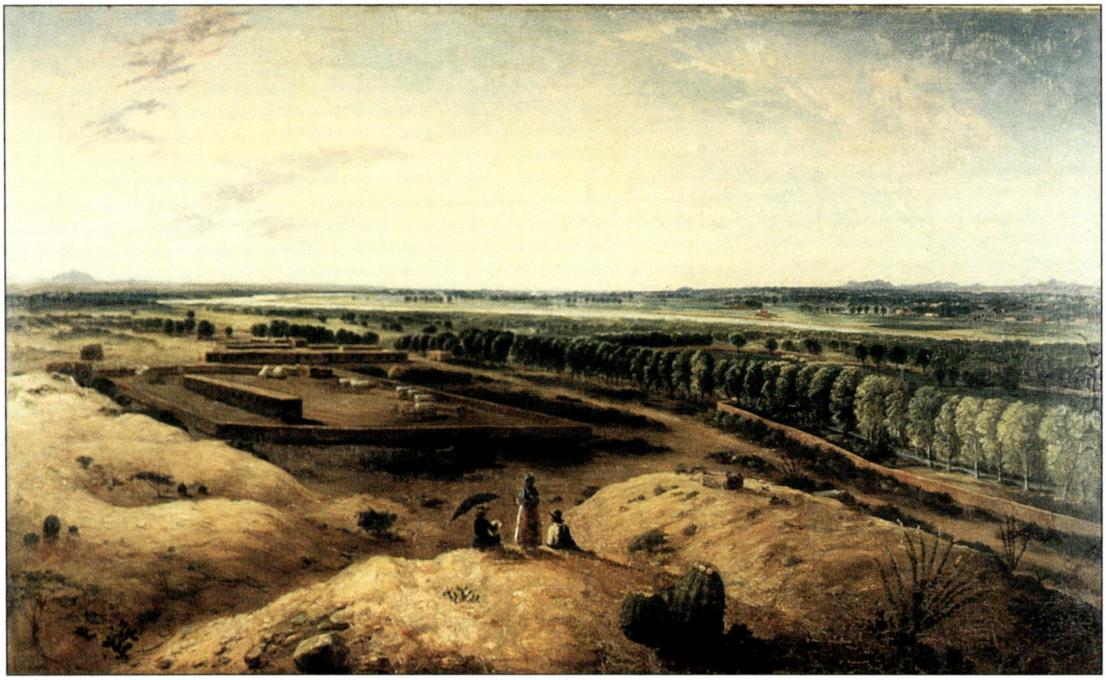

Plate 7

Henry Cheever Pratt

Coons Rancho, Paseo del Norte, Texas, 1851-1853

32" x 50" • Oil on canvas

Loan courtesy of
Texas Memorial Museum,
University of Texas, Austin, Texas

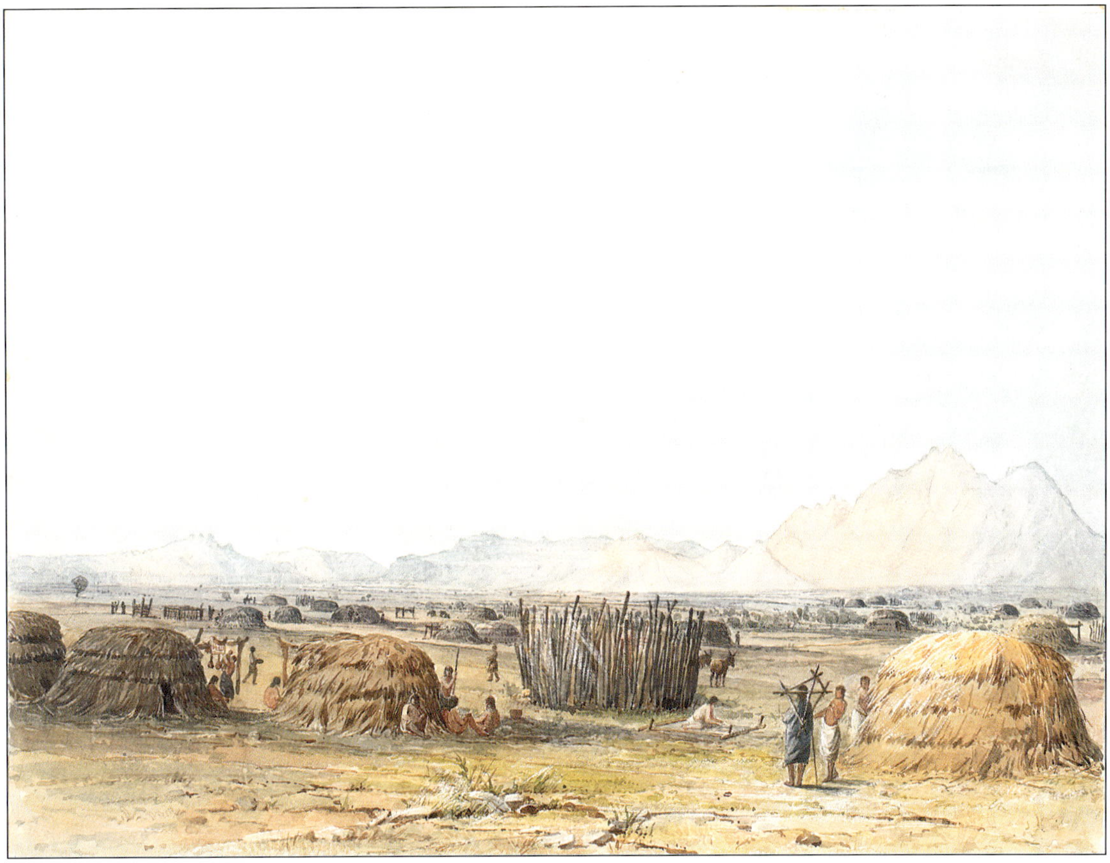

Plate 8

Seth Eastman

Village of the Pimo [sic] *Indians, Gila River*, ca. 1853

9 1/8" x 12 1/8" • Watercolor and pencil

Loan courtesy of Museum of Art, Rhode Island School of Design, Providence [47.112.11]

Photo: Del Bogart

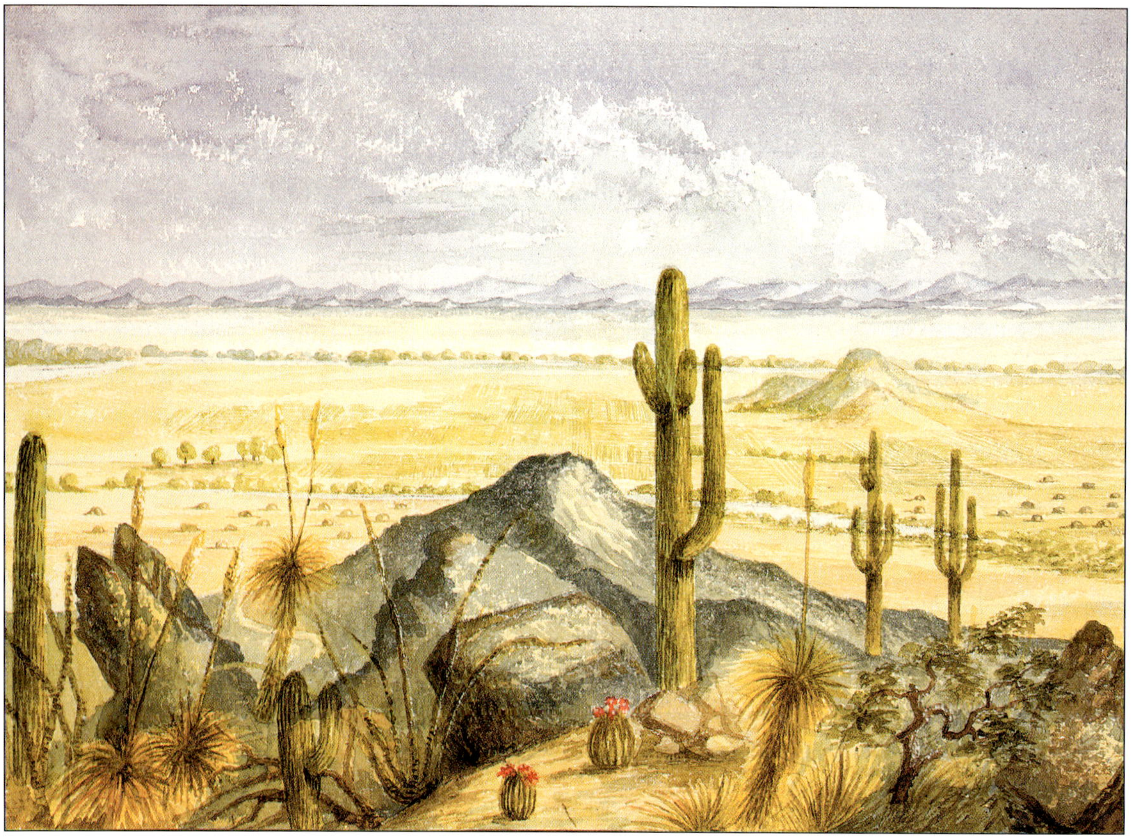

Plate 9
Henry Cheever Pratt
Arizona View Overlooking the Pimo [sic] *Villages*, July 1852
10" x 14" • Watercolor
The John Carter Brown Library, Brown University, Providence, Rhode Island [9]
Photo: Damian Andrus

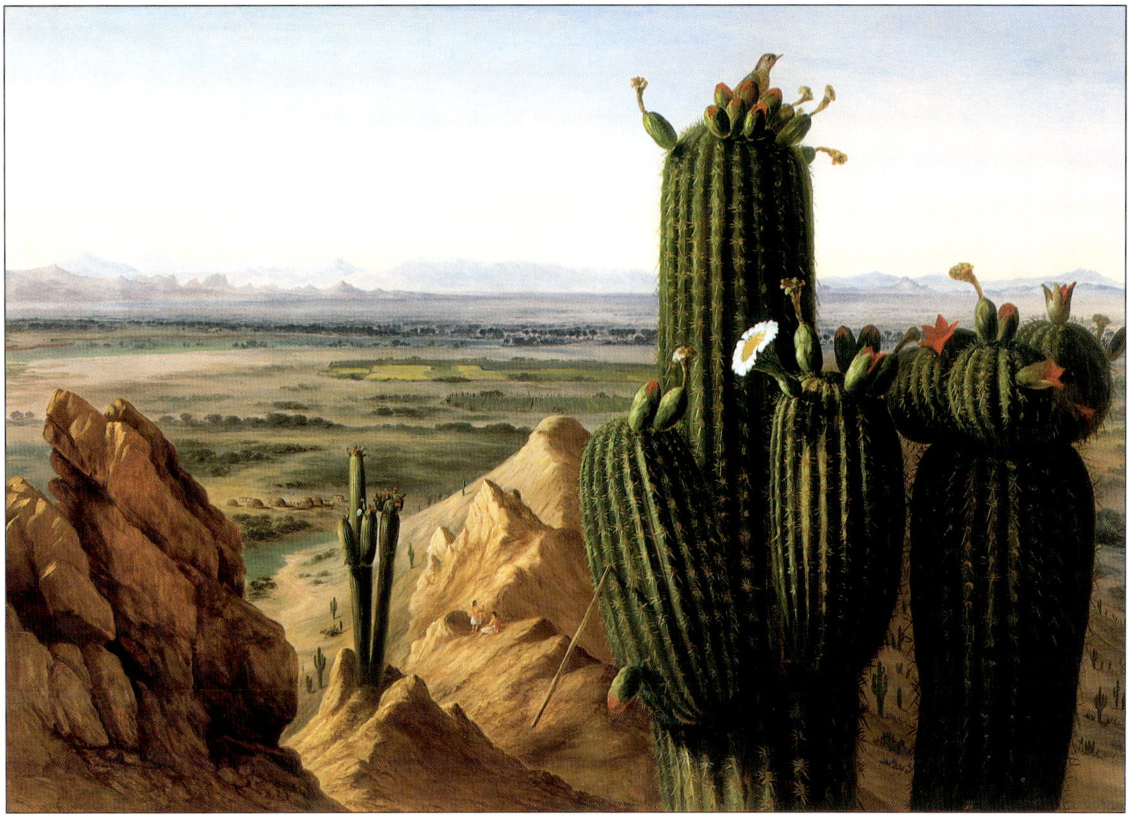

Plate 10

Henry Cheever Pratt

View from Maricopa Mountain near the Rio Gila, 1855

33" x 48" • Oil on canvas

Loan courtesy of The Rosewood Corporation, Dallas, Texas

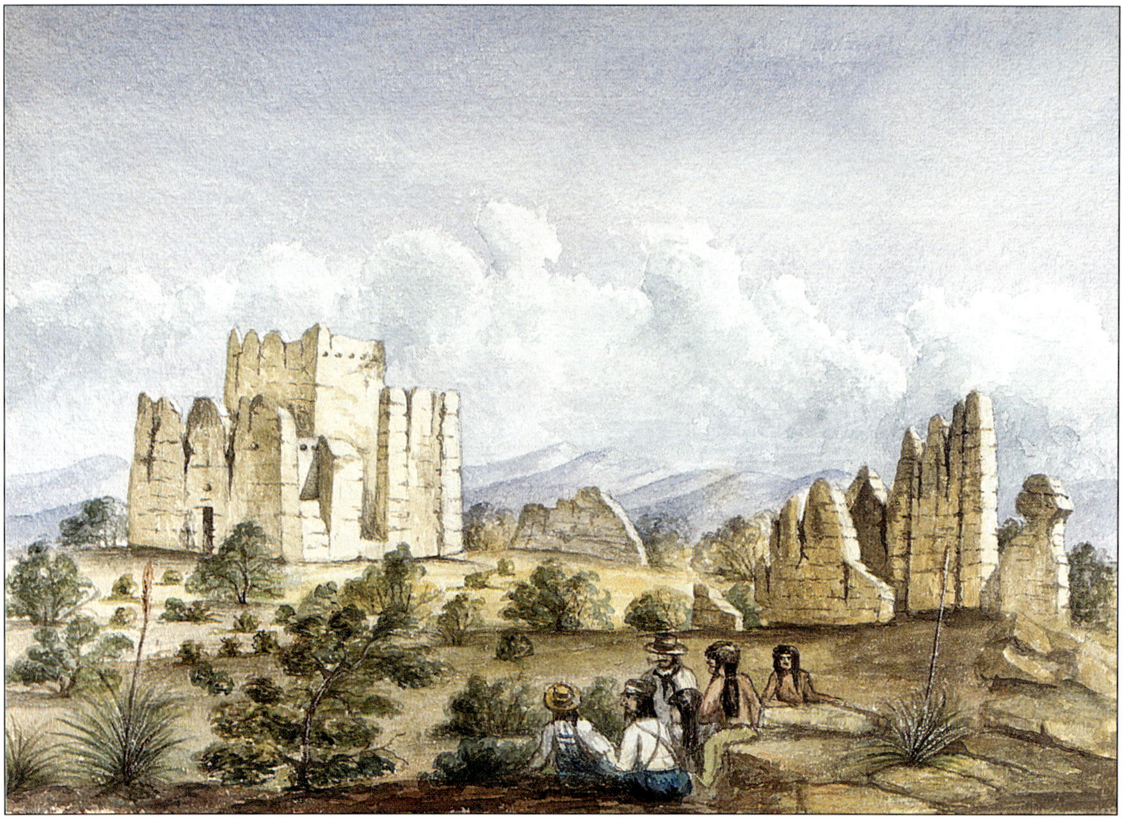

Plate 11

Henry Cheever Pratt

Ruins of Casas Grandes near the River, Gila, Arizona, July 12, 1852

10" x 14" • Watercolor

The John Carter Brown Library, Brown University, Providence, Rhode Island [10]

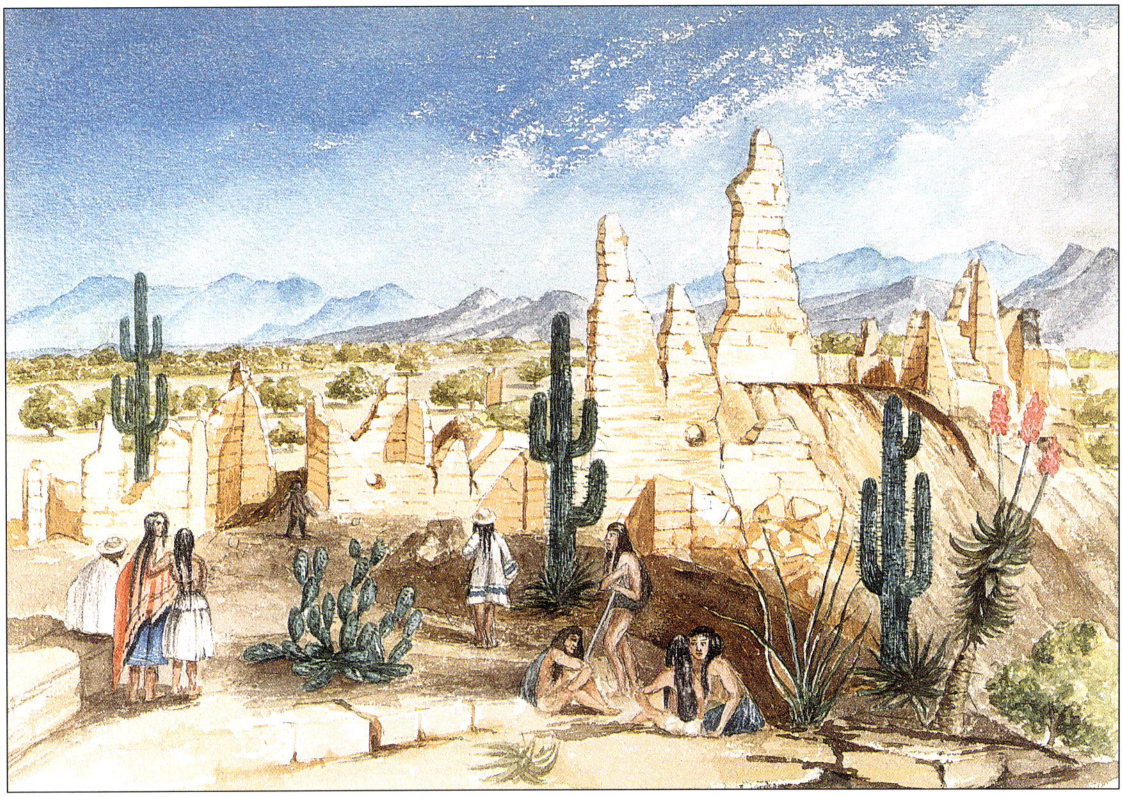

Plate 12

Henry Cheever Pratt

Ruins at Casas Grandes, Chihuahua, Mexico, August 9-10, 1852

9 3/4" x 14" • Watercolor

The John Carter Brown Library, Brown University, Providence, Rhode Island [11]

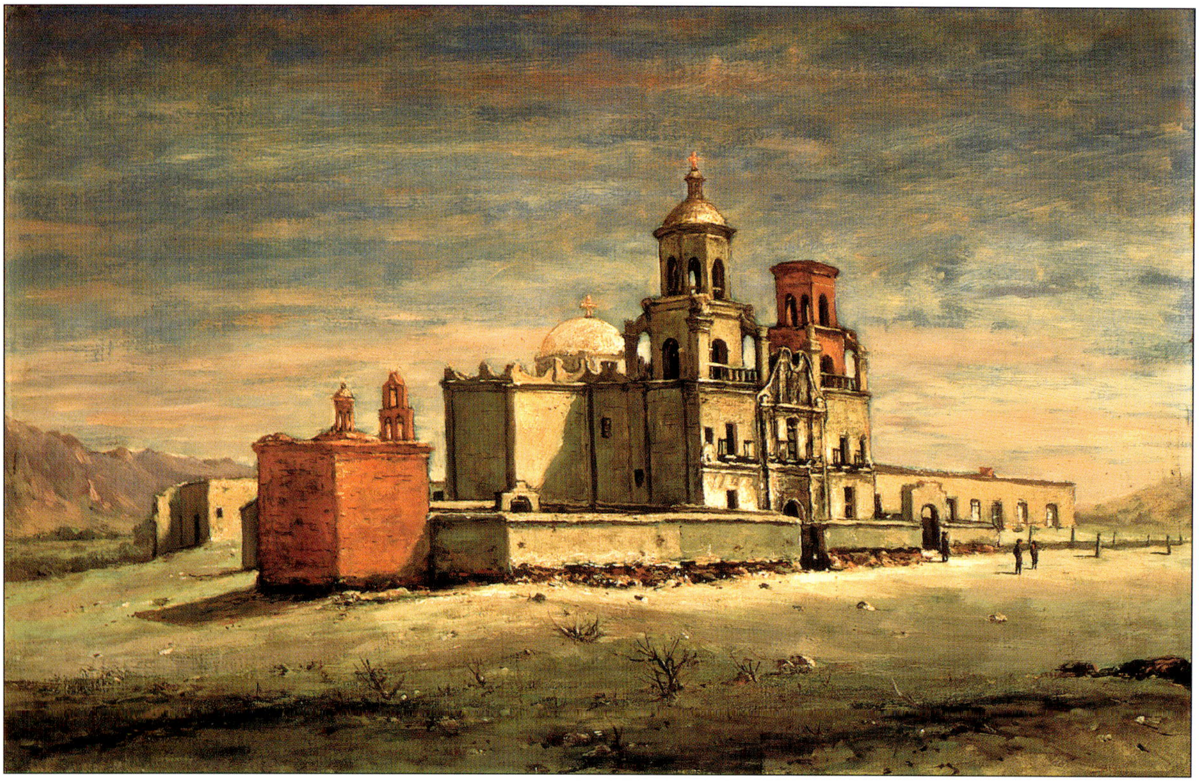

Plate 13
Attributed to Henry Cheever Pratt
San Xavier del Bac Mission Church, after 1860
25" x 29" • Oil on canvas
Loan courtesy of Arthur J. Phelan, Jr.

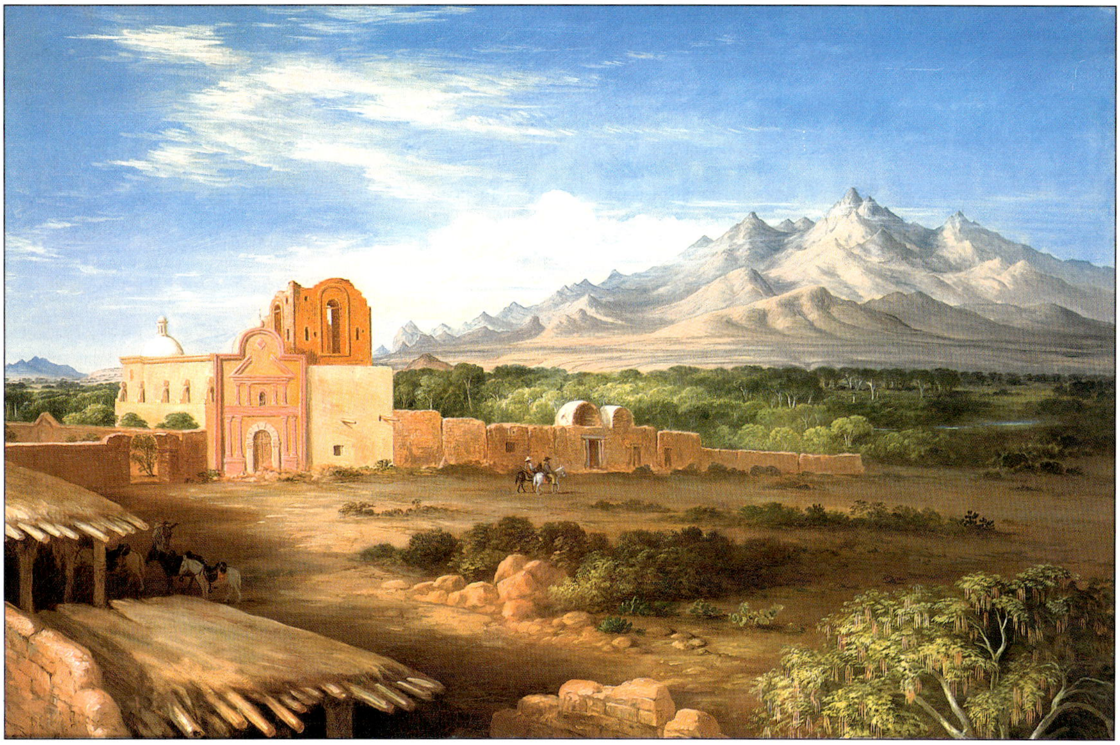

Plate 14

Henry Cheever Pratt

The Deserted Mission of Tumacacori, ca 1854-1855

30" x 46" • Oil on canvas

Loan courtesy of Don Heald
New York, New York

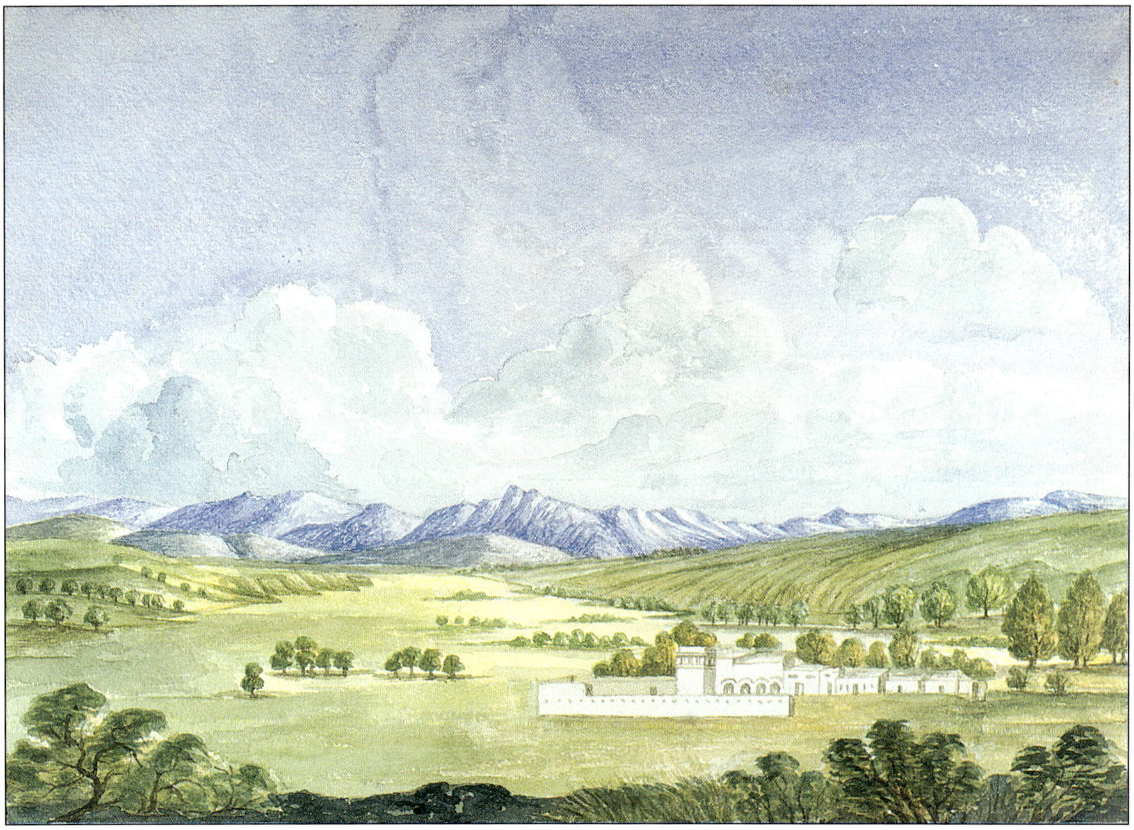

Plate 15

Henry Cheever Pratt

San Lazaro, Arizona,
September 29, 1851

9 3/4" x 14" • Watercolor

The John Carter Brown Library, Brown University, Providence, Rhode Island [21]

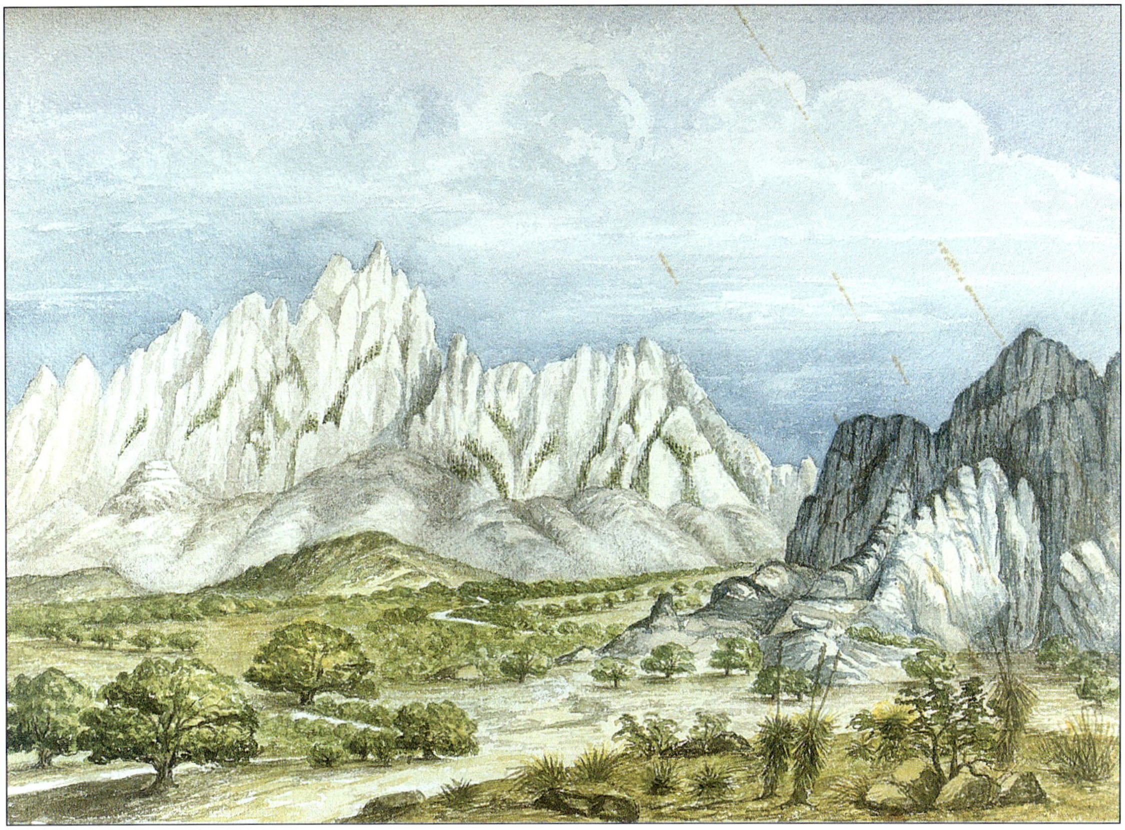

Plate 16

Henry Cheever Pratt

Organ Mountains, New Mexico,
September 1852

10" x 14" • Watercolor

The John Carter Brown Library, Brown University, Providence, Rhode Island [12]

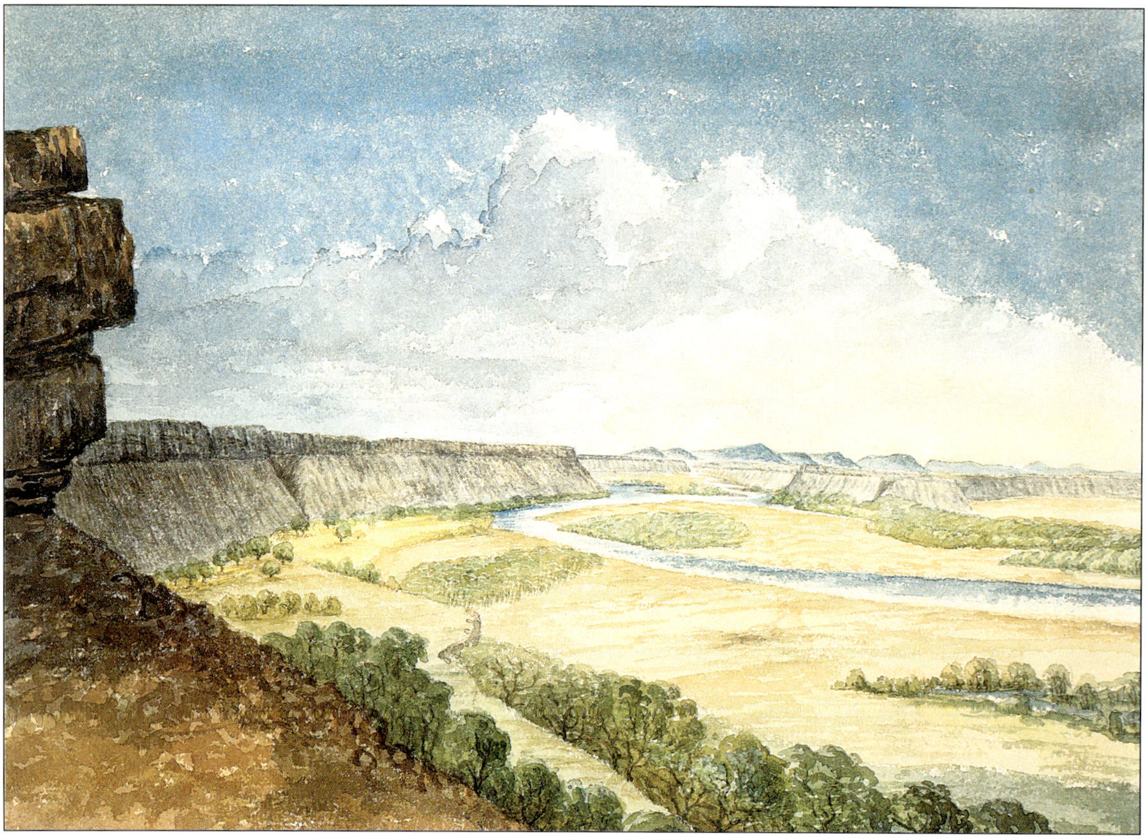

Plate 17

Henry Cheever Pratt

View on the Gila River, Arizona,
July 26-27, 1852

14" x 10" • Watercolor

The John Carter Brown Library, Brown University, Providence, Rhode Island [8]

Photo: Damian Andrus

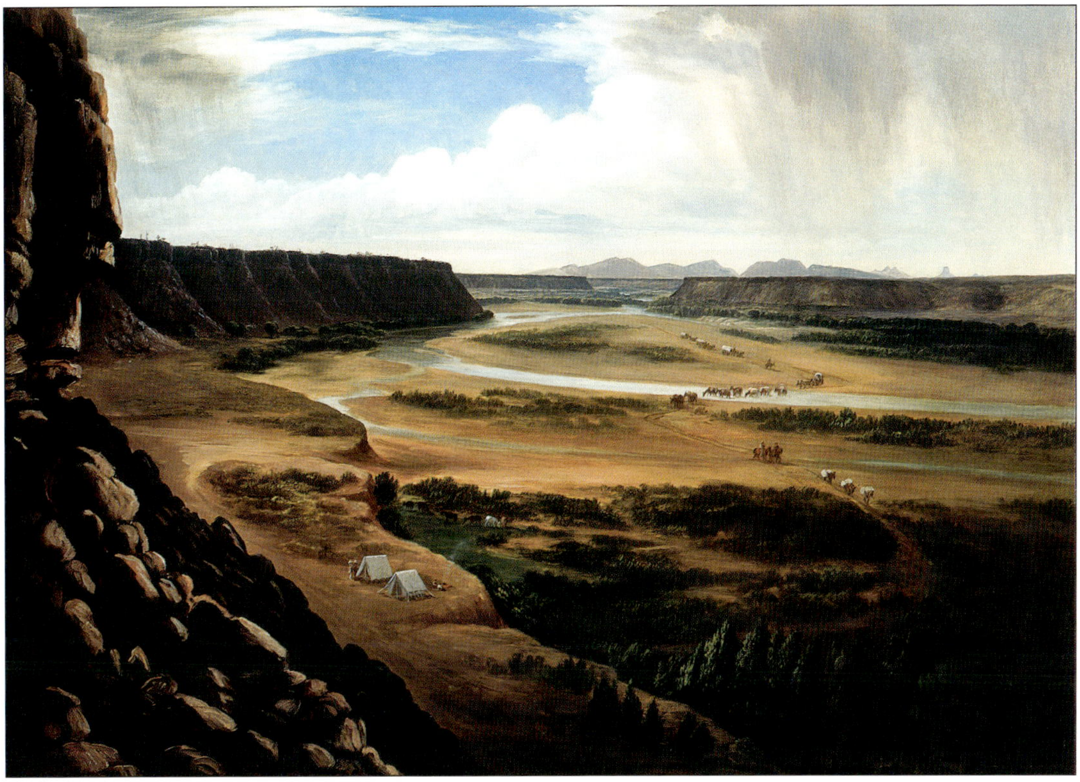

Plate 18

Henry Cheever Pratt

Basin of the Rio Gila, 1855

26 1/4" x 36 1/16" • Oil on canvas

Loan courtesy of Phoenix Art Museum from the Carl S. Dentzel Family Collection, Phoenix, Arizona

Photo: Craig Smith

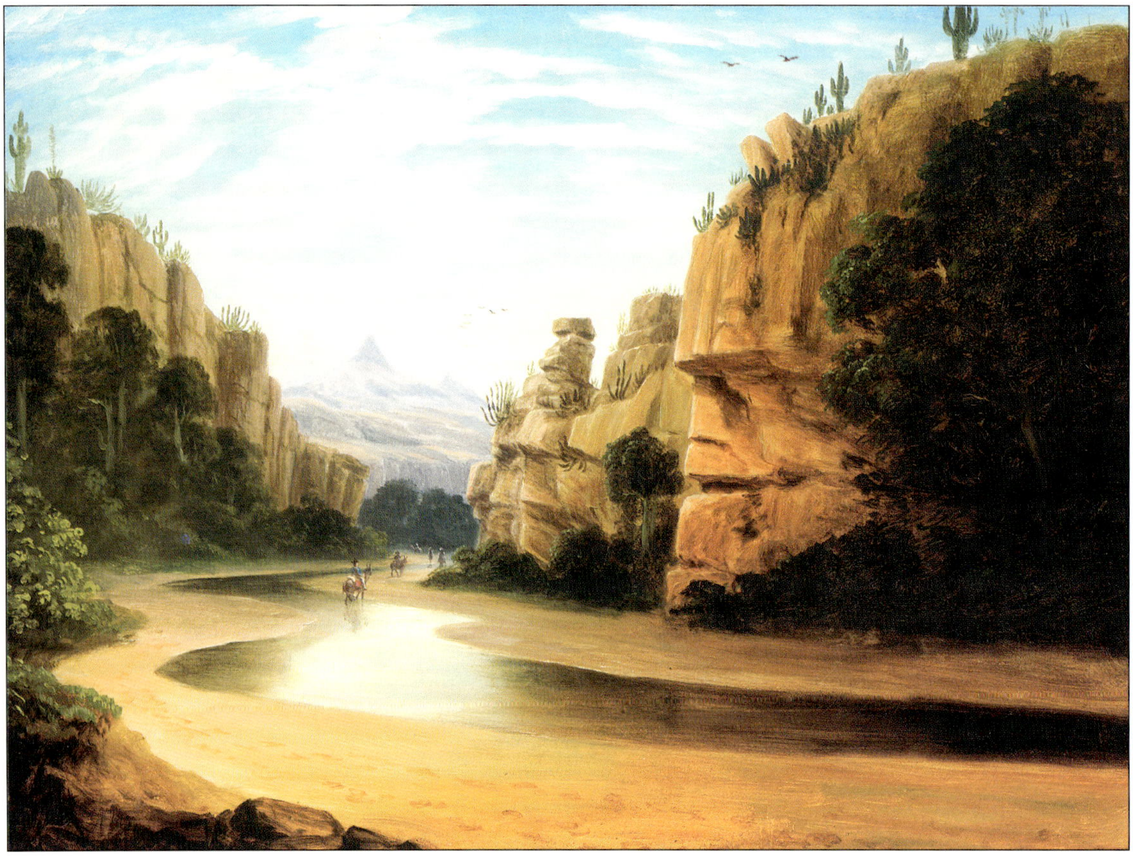

Plate 19

Henry Cheever Pratt

Bartlett Party Traversing a Canyon Bewteen Bacuachi and Arispe, 1855

18" x 24" • Oil on canvas

Loan courtesy of Phoenix Art Museum from the Carl S. Dentzel Family Collection, Phoenix, Arizona

Photo: Craig Smith

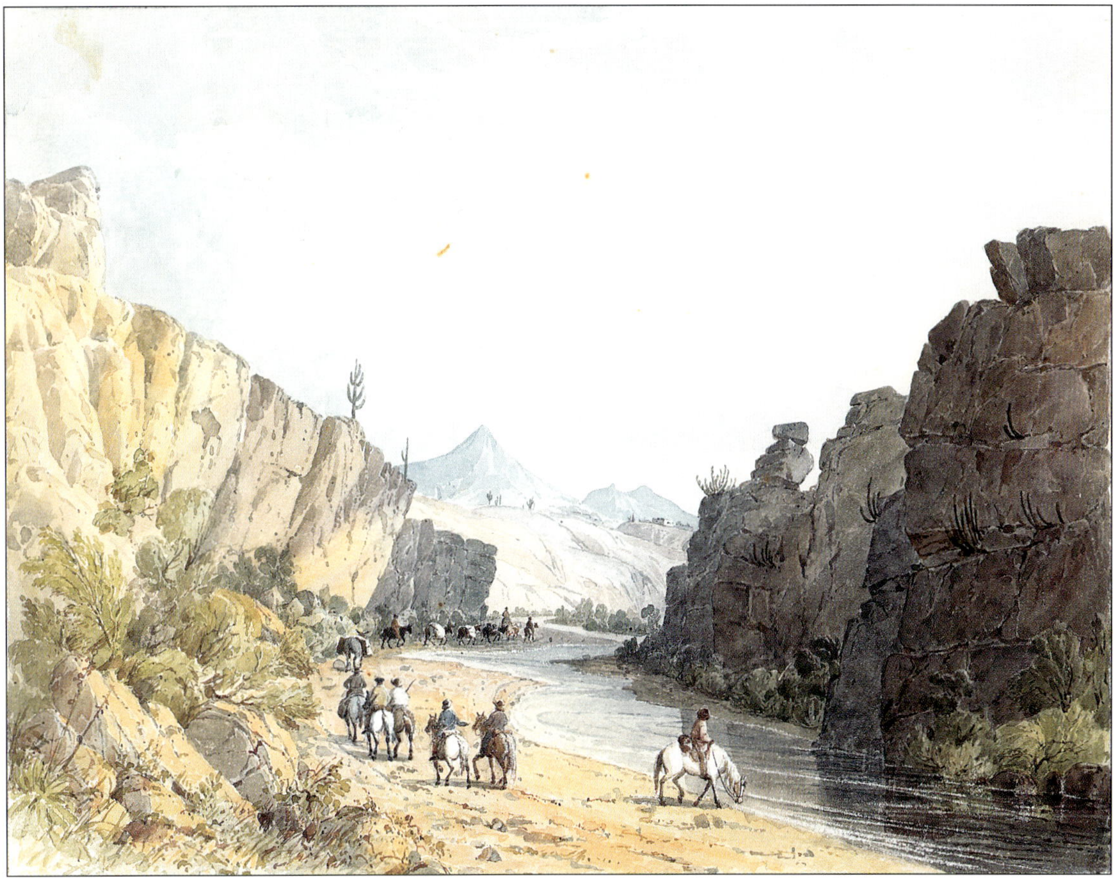

Plate 20

Seth Eastman

Cañon Leading to Magdalena, Sonora, Mexico, ca. 1853

9 1/2" x 12 1/4" • Watercolor and pencil

Loan courtesy of Museum of Art, Rhode Island School of Design, Providence [47.112.1]

Photo: Del Bogart

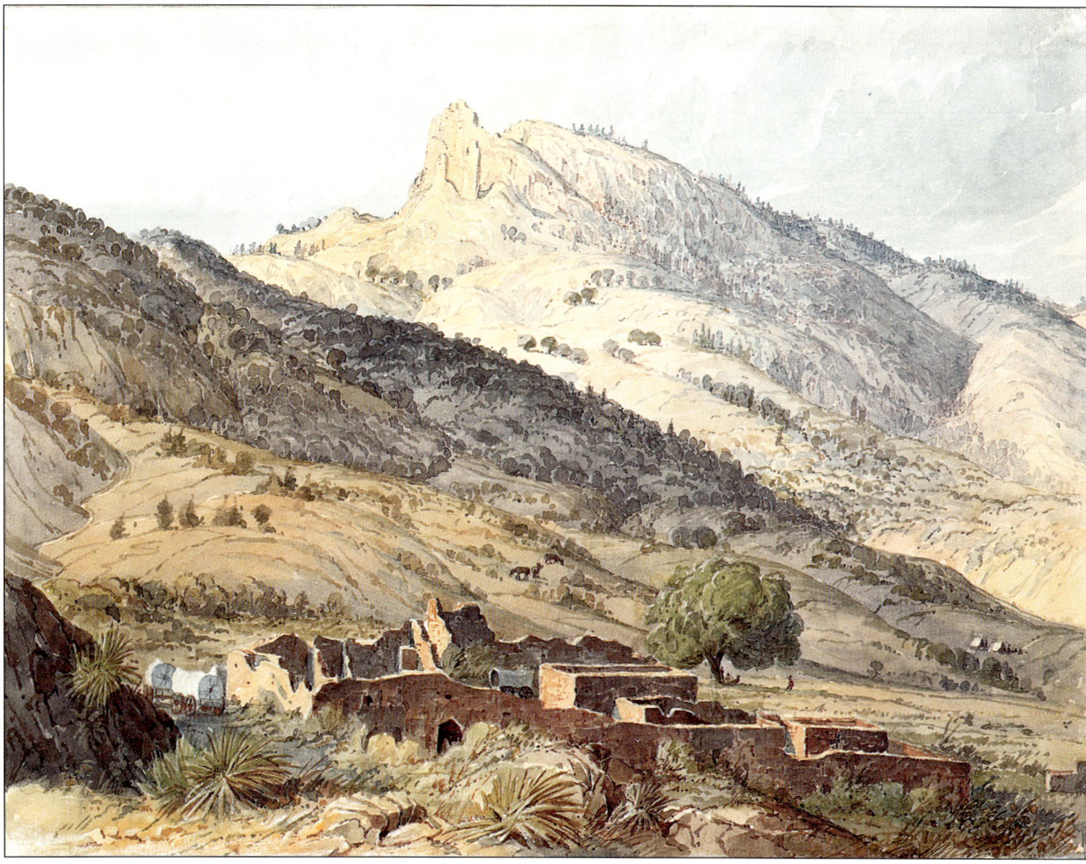

Plate 21

Seth Eastman

Bufa del Cobra [sic], *Rocky Mountains*, ca. 1853

9 1/4" x 12 1/4" • Watercolor and pencil

Loan courtesy of Museum of Art, Rhode Island School of Design, Providence [47.112.3]

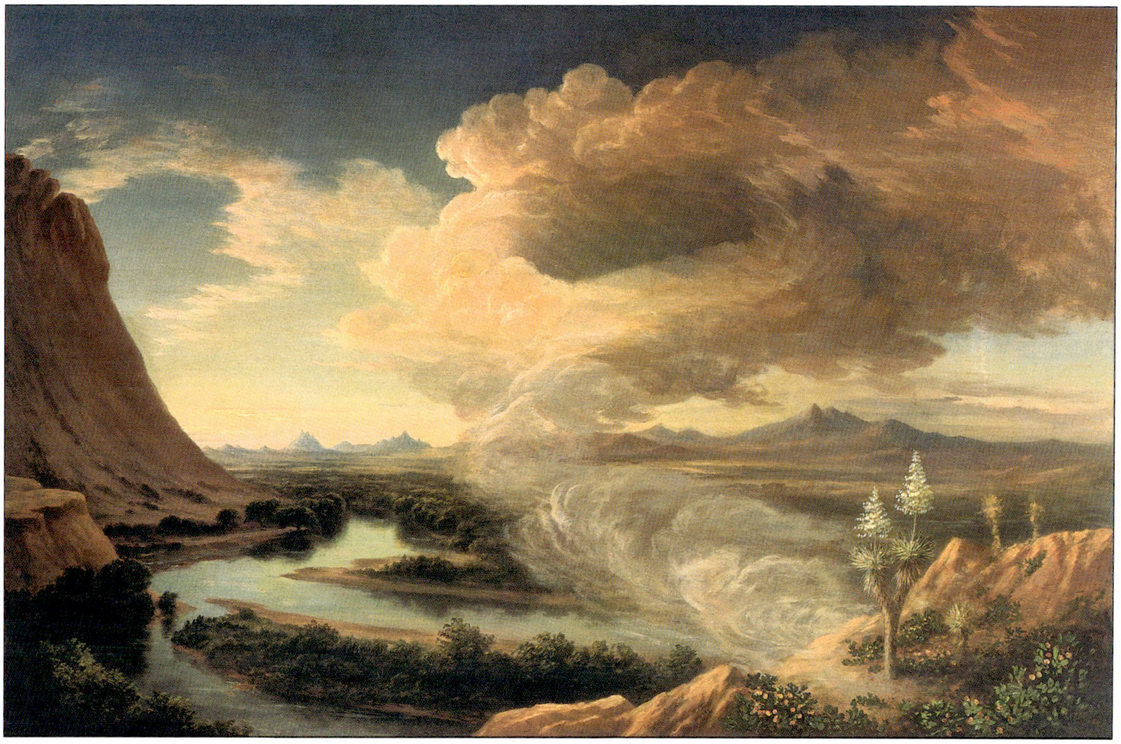

Plate 22

Henry Cheever Pratt

Allegorical Design Representing Evening with a View from Nature Painted on the Gila River, Arizona, ca. 1855

36" x 46" • Oil on canvas

Loan courtesy Collection of Philip Y. De Normandie

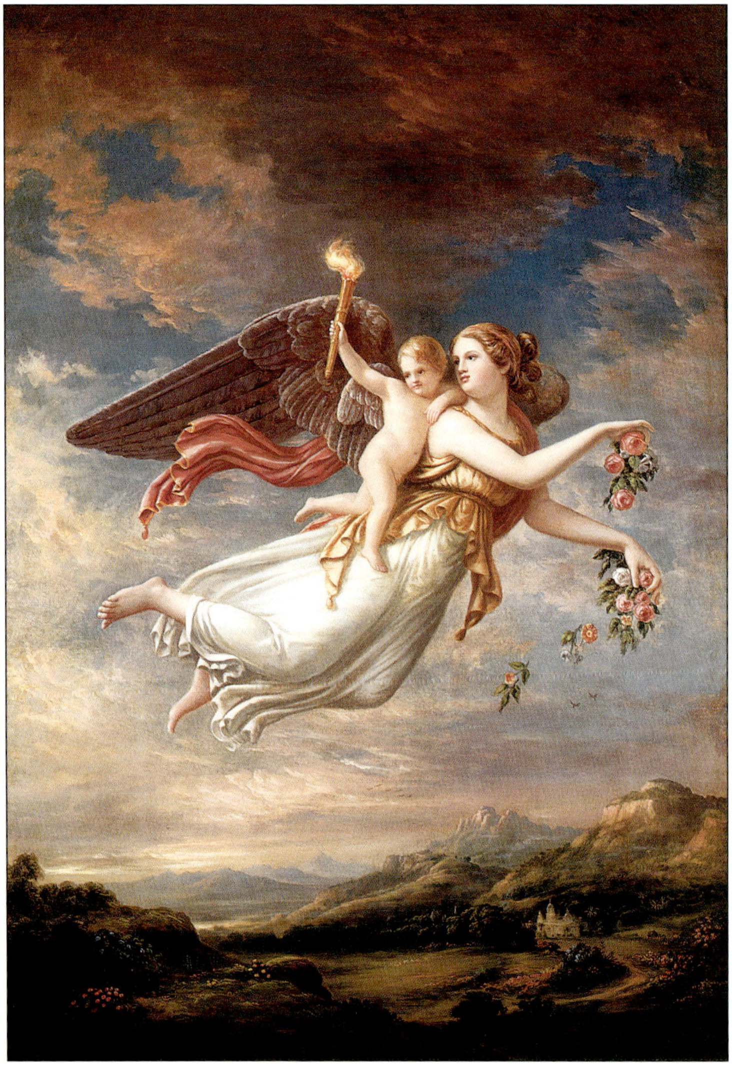

Plate 23
Henry Cheever Pratt
Dawn, 1862
58" x 41" • Oil on canvas
Loan courtesy of
Rhoda and David T. Chase

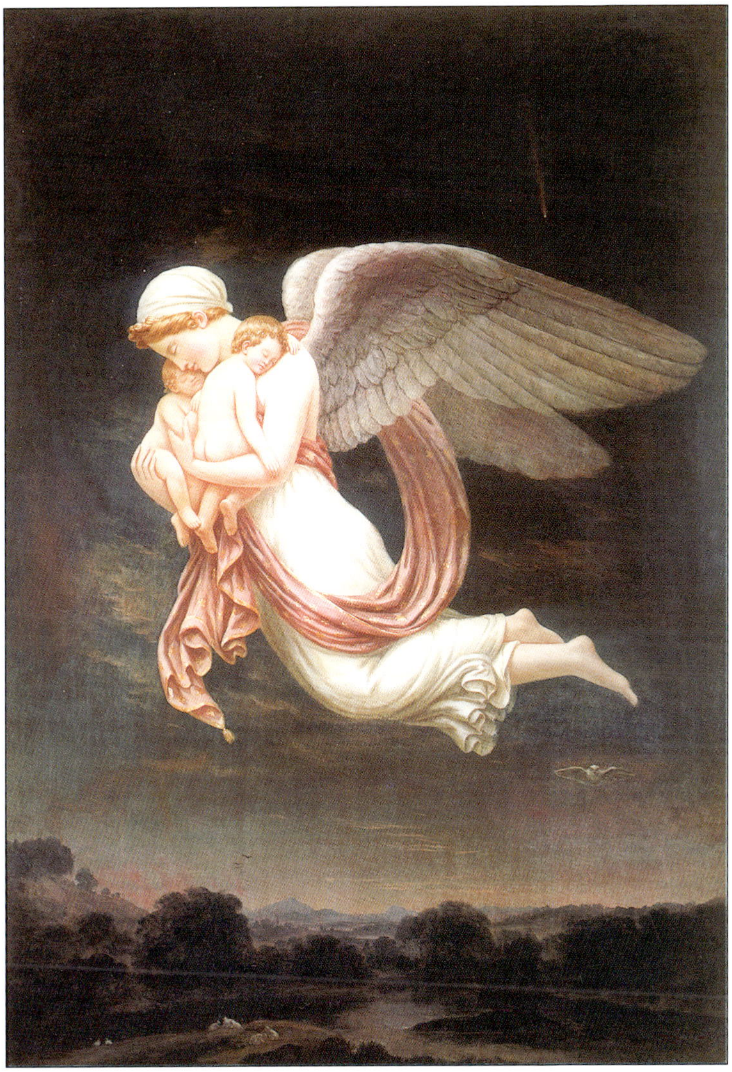

Plate 24
Henry Cheever Pratt
Night, 1862
58" x 41" • Oil on canvas
Loan courtesy of
Rhoda and David T. Chase

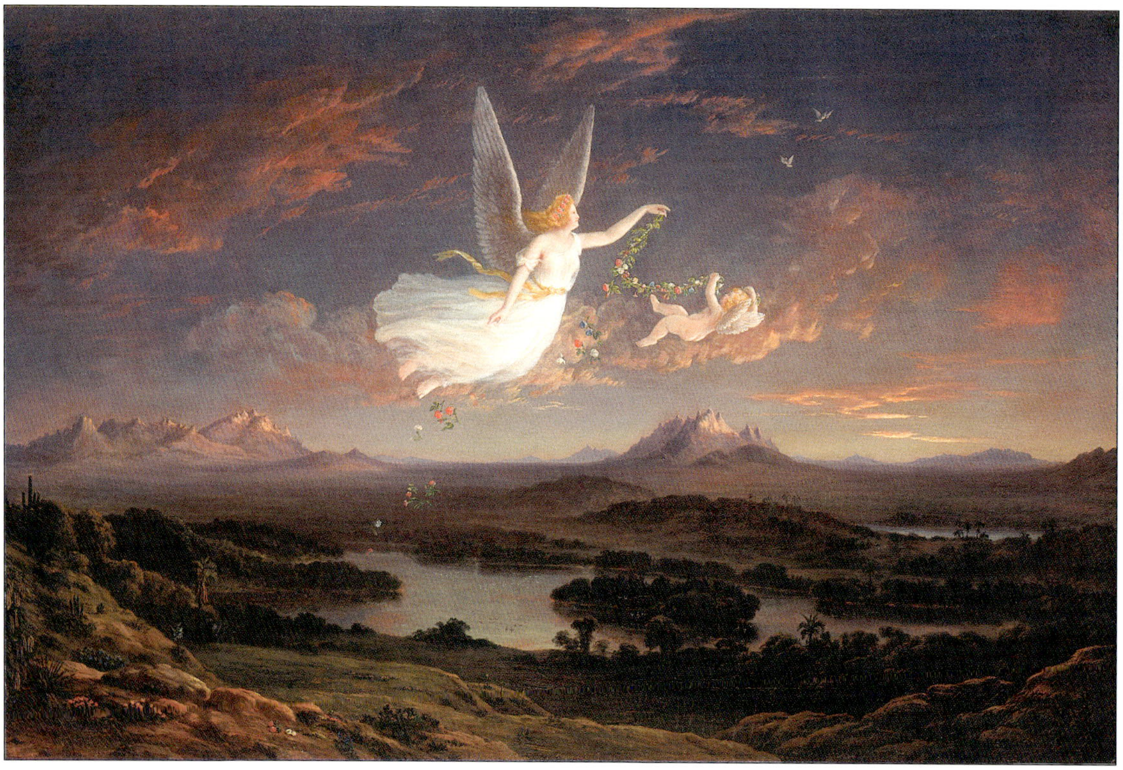

Plate 25
Henry Cheever Pratt
Evening Scene on the Gila River, Arizona, 1862
30" x 46" • Oil on canvas
Loan courtesy Collection of Philip Y. De Normandie

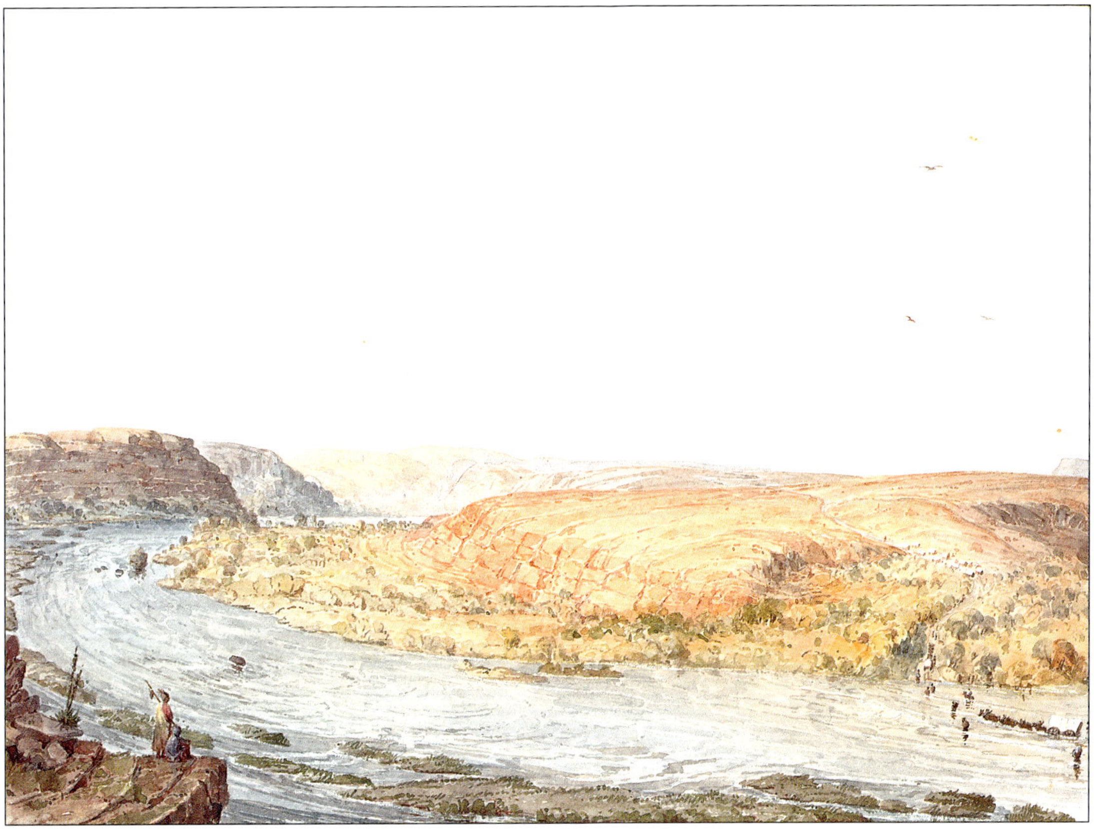

Plate 26

Seth Eastman

Fording Devil's River, Texas, ca. 1853

8 3/4" x 12" • Watercolor and pencil

Loan courtesy of Museum of Art, Rhode Island School of Design, Providence [47.112.10]

Photo: Del Bogart

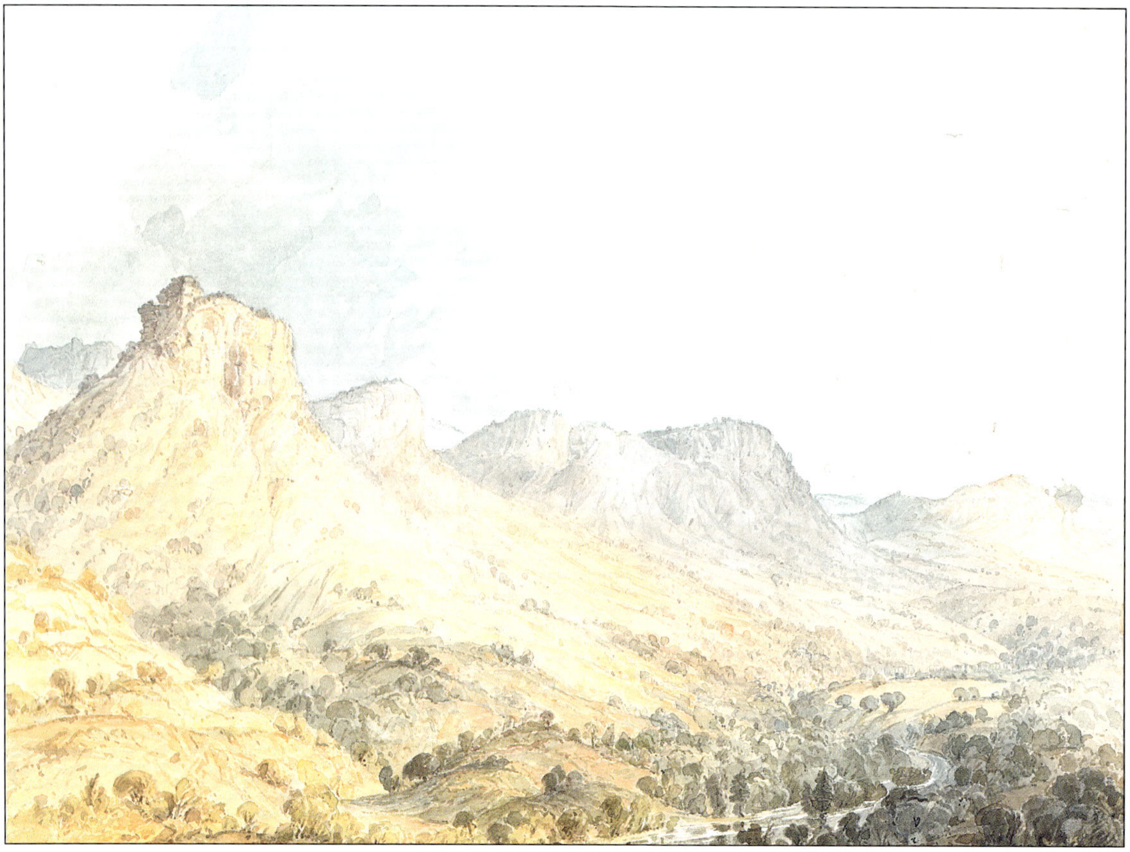

Plate 27
Seth Eastman
Cañon Leading to the Copper Mines at Santa Rita del Cobra [sic], ca. 1853
9 3/8" x 12 1/2" • Watercolor and pencil
Loan courtesy of Museum of Art, Rhode Island School of Design, Providence [47.112.12]
Photo: Del Bogart

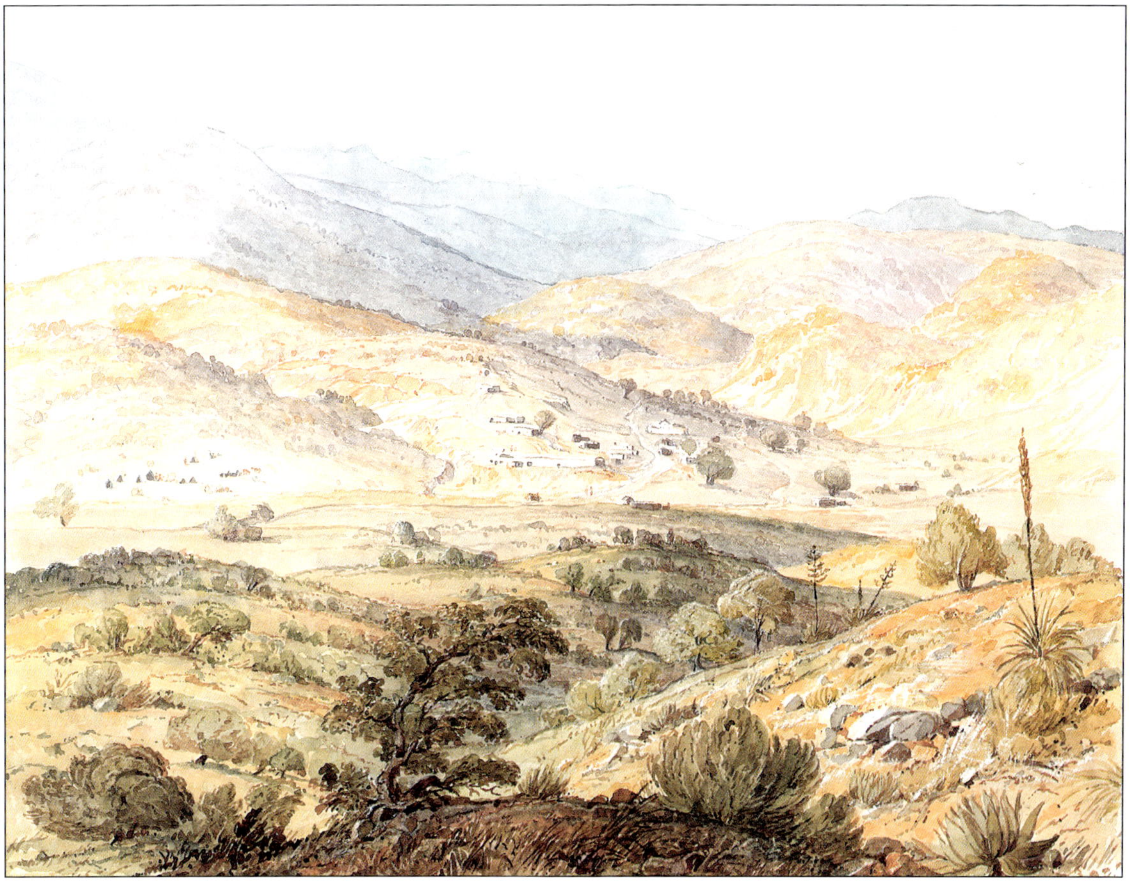

Plate 28

Seth Eastman

Santa Rita del Cobra [sic], *Rocky Mountains, New Mexico*, ca. 1853

9 1/2" x 12 1/2" • Watercolor and pencil

Loan courtesy of Museum of Art, Rhode Island School of Design, Providence [47.112.6]

Photo: Del Bogart

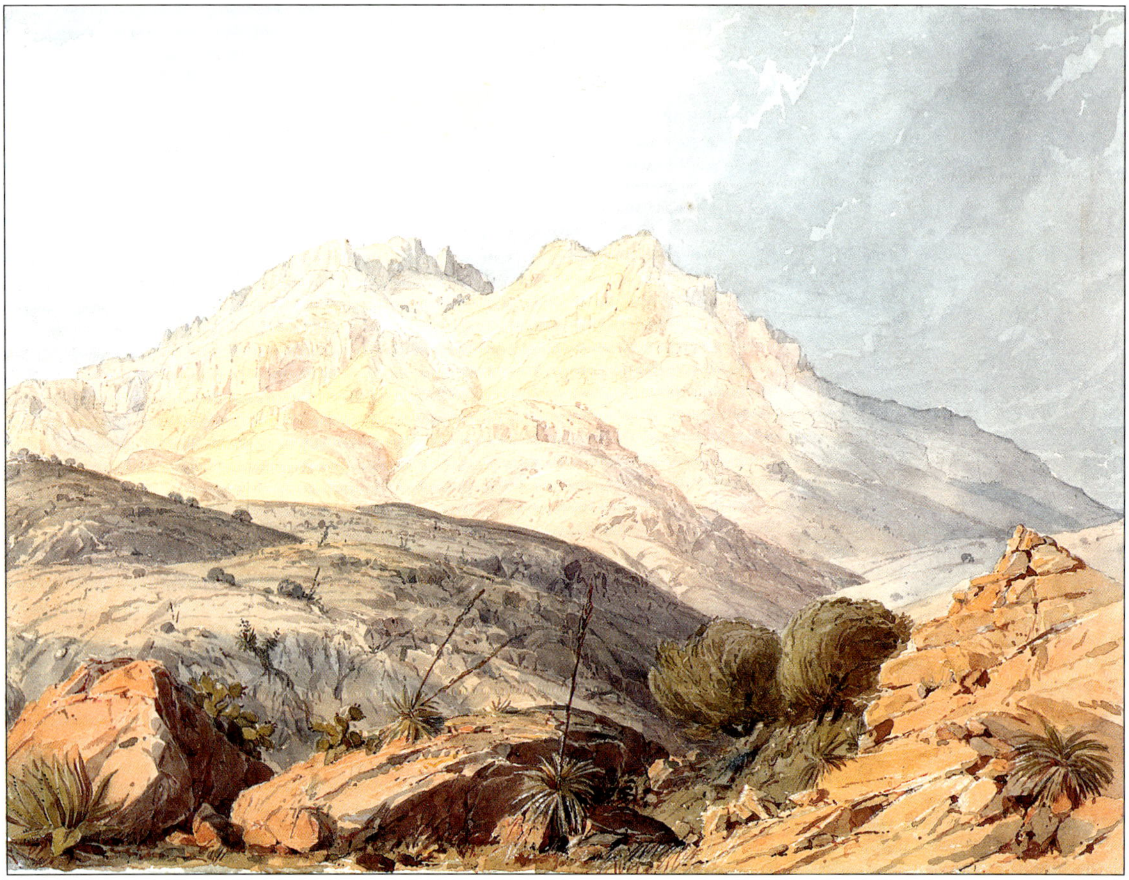

Plate 29

Seth Eastman

View from the Camp near the Boundary Line 4th day from Fort Webster, ca. 1853

9 1/2" x 12 1/2" • Watercolor and Pencil

Loan courtesy of Museum of Art, Rhode Island School of Design, Providence [47.112.4]

Photo: Del Bogart

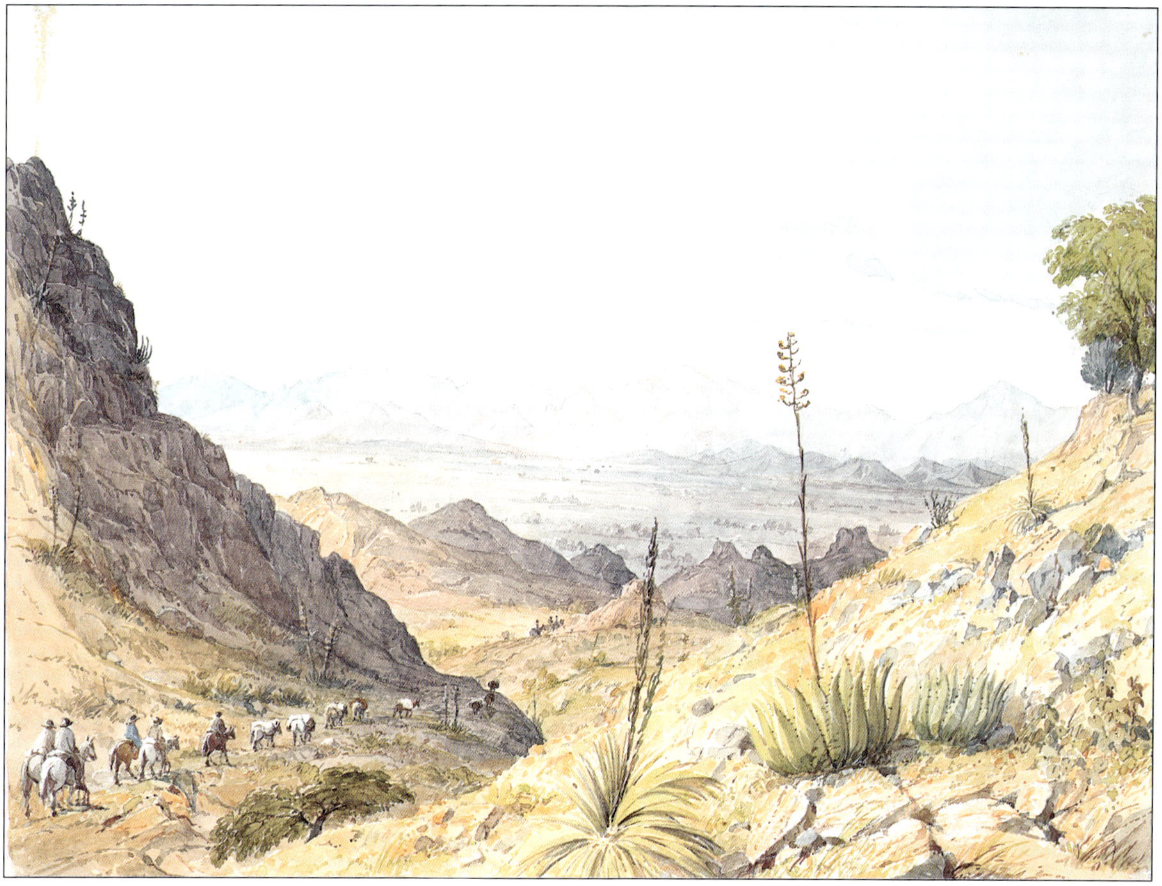

Plate 30

Seth Eastman

Pass at the Pitoncillo, [sic] *Chiricahui Mountains in the Distance,* ca. 1853

9 1/4" x 12 3/8" • Watercolor and pencil

Loan courtesy of Museum of Art, Rhode Island School of Design, Providence [47.112.5]

Photo: Del Bogart

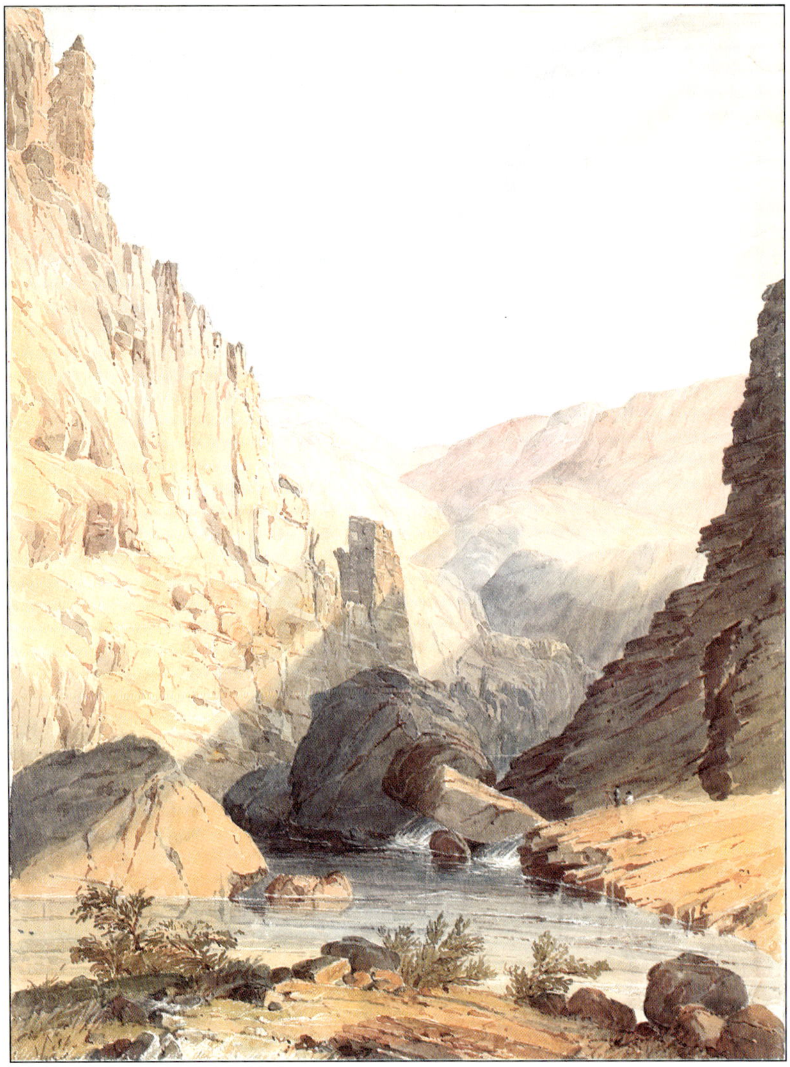

Plate 31
Seth Eastman
Great Cañon, Rio Gila, ca. 1853
12 1/2" x 9 5/8" • Watercolor and pencil
Loan courtesy of Museum of Art, Rhode Island School of Design, Providence [47.112.9]
Photo: Del Bogart

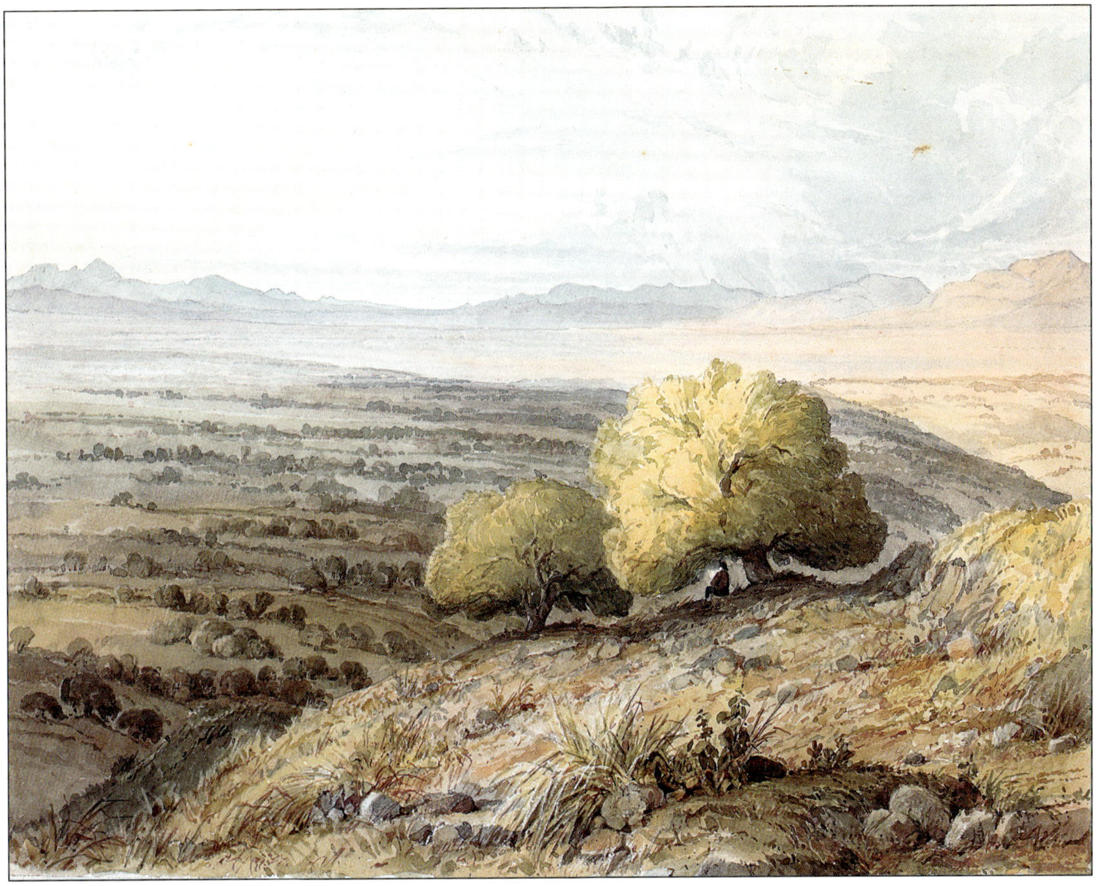

Plate 32

Seth Eastman

Santa Cruz Valley, ca. 1853

9 3/8" x 12 1/4" • Watercolor and pencil

Loan courtesy of Museum of Art, Rhode Island School of Design, Providence [47.112.13]

Photo: Del Bogart

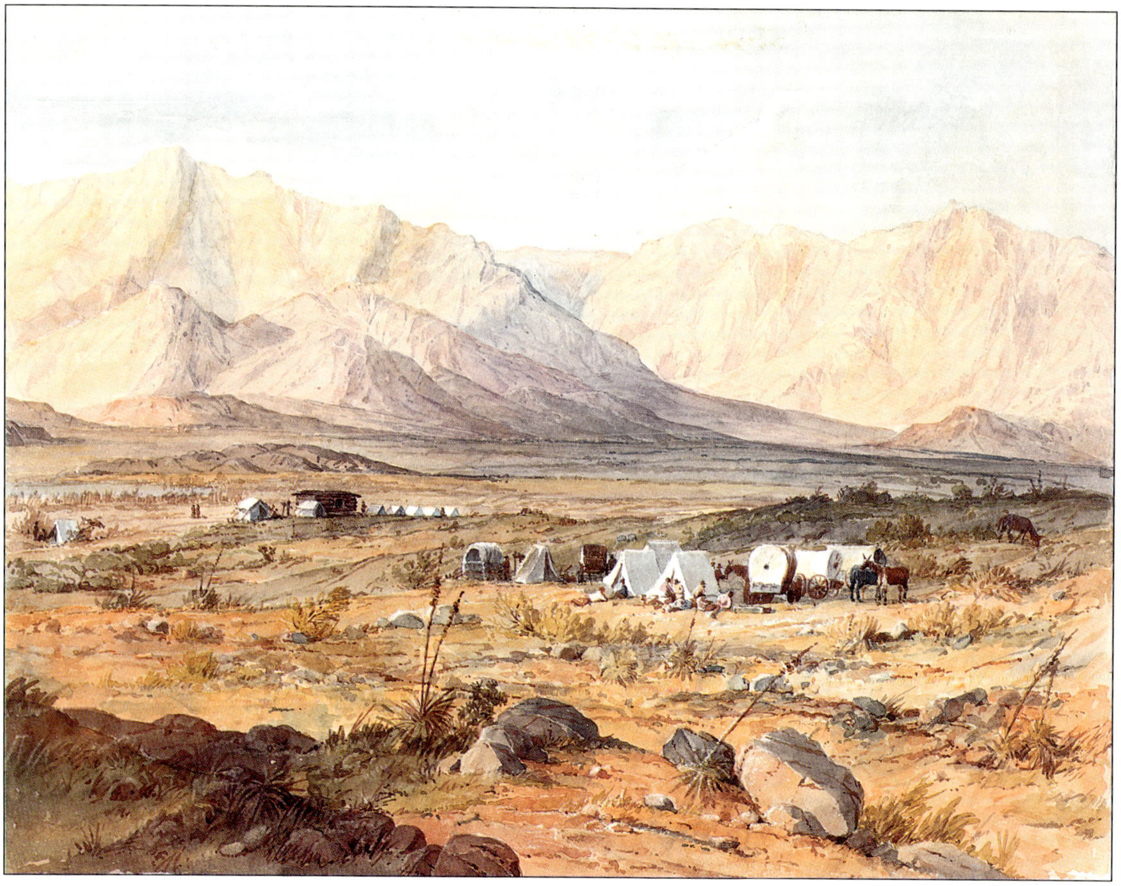

Plate 33
Seth Eastman
Vallecito, California, ca. 1853
9 3/8" x 12 3/8" • Watercolor and pencil
Loan courtesy of Museum of Art, Rhode Island School of Design, Providence [47.112.7]
Photo: Del Bogart

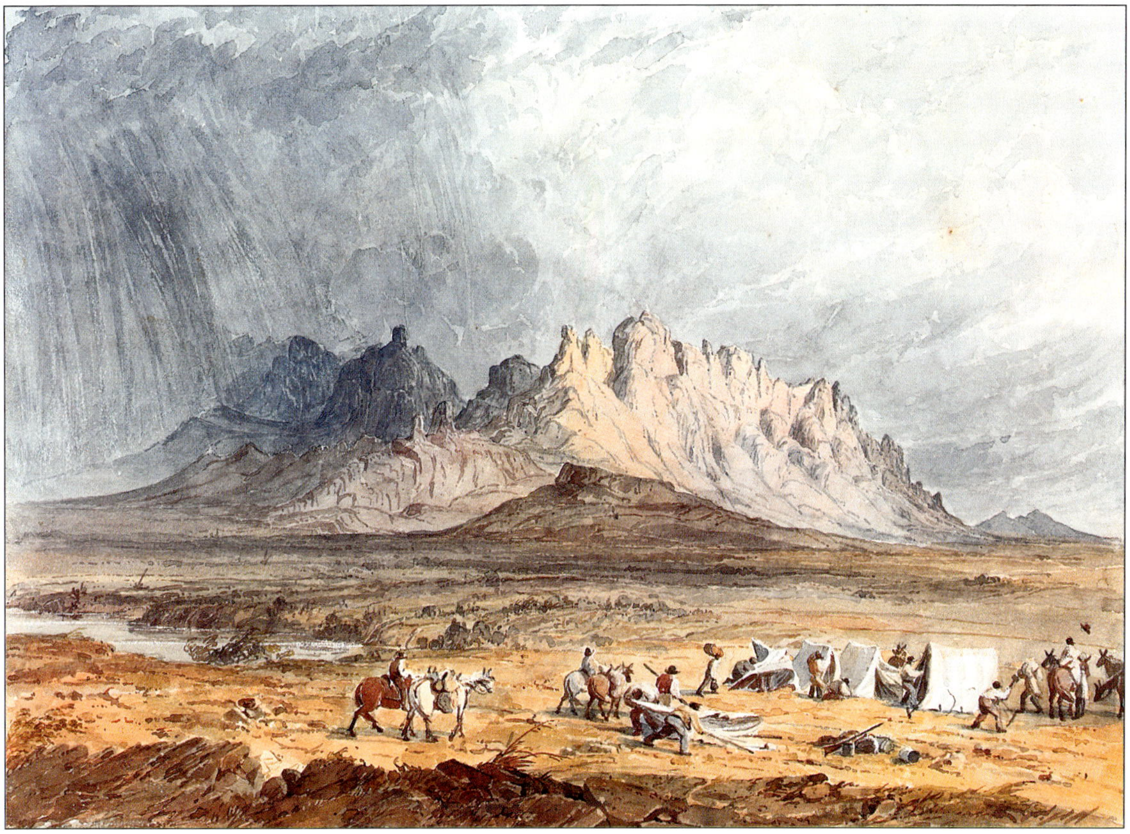

Plate 34

Seth Eastman

Camping in a Storm on the Mexican Plateau, ca. 1853

9 1/8" x 12 1/2" • Watercolor and pencil

Loan courtesy of Museum of Art, Rhode Island School of Design, Providence [47.112.2]

Photo: Del Bogart

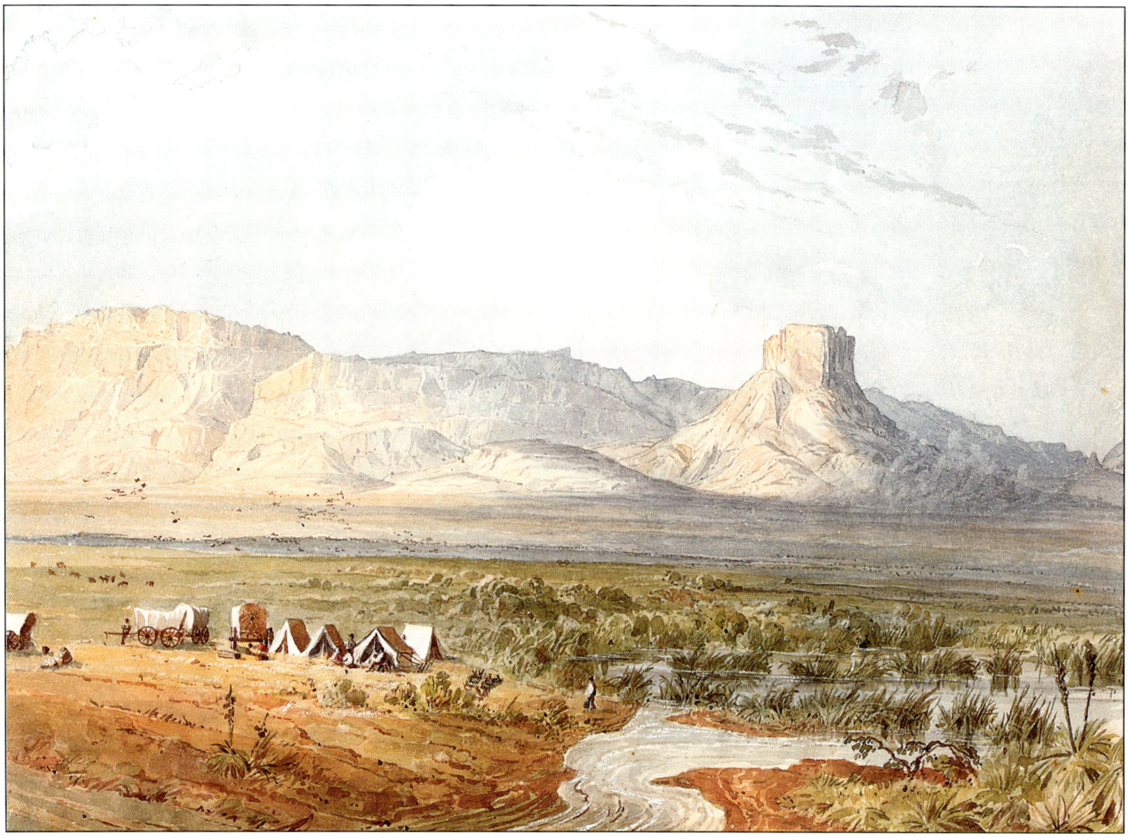

Plate 35
Seth Eastman
Camp near Lake de los Patos, Chihuahua, ca. 1853
9" x 12 3/8" • Watercolor and pencil
Loan courtesy of Museum of Art, Rhode Island School of Design, Providence [47.112.8]
Photo: Del Bogart

CHECK LIST OF THE EXHIBITION

The works on this list are in order by lender.

The reproductions in this section are works that are not reproduced elsewhere in the body of this book.

1 Henry Cheever Pratt (Plate 6 • Pg 115)
Portrait of Benjamin Franklin Coons, 1851
30" x 37" • Oil on canvas
Loan courtesy of Philip and Amy Alfeld

2 Henry Cheever Pratt (Plate 2 • Pg 111)
Portrait of John Russell Bartlett,
October 1, 1852
36 1/4" x 29 1/4" • Oil on canvas
Loan courtesy of Amon Carter Museum,
Ft. Worth, Texas

3 Henry Cheever Pratt (Plate 23 • Pg 132)
Dawn, 1862
58" x 41" • Oil on canvas
Loan courtesy of Rhoda and
David T. Chase

4 Henry Cheever Pratt (Plate 24 • Pg 133)
Night, 1862
58" x 41" • Oil on canvas
Loan courtesy of Rhoda and
David T. Chase

5 Henry Cheever Pratt (Plate 1 • Pg 110)
Indians Bathing: A Scene of New England,
1843
27 1/2" x 37 1/2" • Oil on canvas
Loan courtesy of William Blundin,
New York, New York

6 Henry Cheever Pratt (Plate 25 • Pg 134)
Evening Scene on the Gila River, 1862
30" x 46" • Oil on canvas
Loan courtesy Collection of
Philip Y. De Normandie

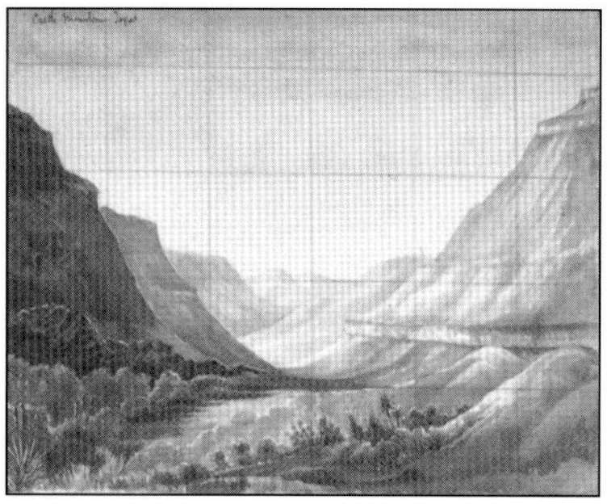

Ex. #8

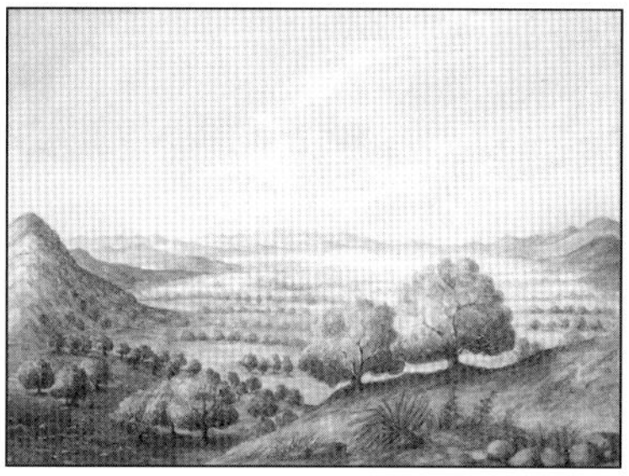

Ex. #10

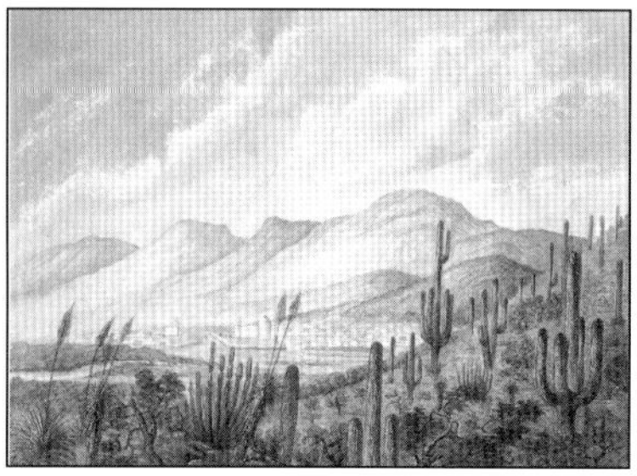

Ex. #11

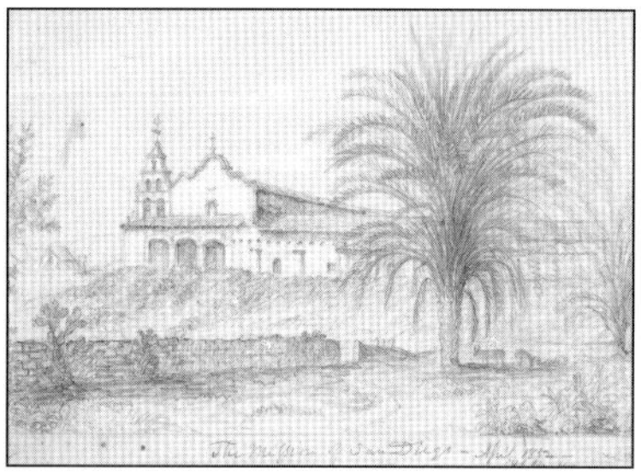

Ex. #12

7 Henry Cheever Pratt (Plate 3 • Pg 112)
Fort Yuma, Colorado River, June 1852
9 3/4" x 9 1/8" • Watercolor
The John Carter Brown Library, Brown University, Providence, Rhode Island [1]

8 Henry Cheever Pratt
Castle Mountain Pass, Texas, ca. 1853
9 1/4" x 11 1/2" • Watercolor
The John Carter Brown Library, Brown University, Providence, Rhode Island [2]

9 John Russell Bartlett (Fig. 20 • Pg 43)
Church at El Paso del Norte, Chihuahua, November 23, 1850
9" x 11 1/4" • Sepia wash
The John Carter Brown Library, Brown University, Providence, Rhode Island [3]

10 Henry Cheever Pratt
Valley Leading to Santa Cruz, Sonora, ca. 1852
9 7/8" x 13 7/8" • Watercolor
The John Carter Brown Library, Brown University, Providence, Rhode Island [4]

11 Henry Cheever Pratt
Magdelena, Sonora, ca. 1852
9 3/4" x 13 7/8" • Watercolor
The John Carter Brown Library, Brown University, Providence, Rhode Island [5]

12 Henry Cheever Pratt
The Mission of San Diego, April 1852
4 1/2" x 7" • Pencil
The John Carter Brown Library, Brown University, Providence, Rhode Island [6]

13 Henry Cheever Pratt (Plate 4 • Pg 113)
Junction of the Colorado and the Gila Rivers, Looking up the Gila, June 16, 1852
9 3/4" x 14" • Watercolor
The John Carter Brown Library, Brown University, Providence, Rhode Island [7]

14 Henry Cheever Pratt
(Fig. 13 • Pg. 39, Plate 17 • Pg 126)
View on the Gila River, Arizona, July 26-27, 1852
14" x 10" • Watercolor
The John Carter Brown Library, Brown University, Providence, Rhode Island [8]

15 Henry Cheever Pratt (Plate 9 • Pg 118)
Arizona View Overlooking the Pimo [sic] *Villages,* July 1852
10" x 14" • Watercolor
The John Carter Brown Library, Brown University, Providence, Rhode Island [9]

16 Henry Cheever Pratt (Plate 11 • Pg 120)
Ruins of Casas Grandes Near the River, Gila, Arizona, July 12, 1852
10" x 14" • Watercolor
The John Carter Brown Library, Brown University, Providence, Rhode Island [10]

17 Henry Cheever Pratt (Plate 12 • Pg 121)
Ruins at Casas Grandes, Chihuahua, Mexico, August 9-10, 1852
9 3/4" x 14" • Watercolor
The John Carter Brown Library, Brown University, Providence, Rhode Island [11]

18 Henry Cheever Pratt (Plate 16 • Pg 125)
Organ Mountains, New Mexico, September 1852
10" x 14" • Watercolor
The John Carter Brown Library, Brown University, Providence, Rhode Island [12]

19 Henry Cheever Pratt (Fig. 19 • Pg 41)
Chimney Rock as seen from Ft. Yuma, River, Colorado, June 16, 1852
3 3/4" x 9" • Pencil
The John Carter Brown Library, Brown University, Providence, Rhode Island [14]

20 Henry Cheever Pratt (Fig. 23 • Pg 43)
Ruins of the Church at San Elizario on the Rio Grande, October 8, 1852
8 1/2" x 6" • Pencil
The John Carter Brown Library, Brown University, Providence, Rhode Island [17]

21 Henry Cheever Pratt (Fig. 32 • Pg 54)
Church of San Xavier del Bac, Sonora, July 19, 1852
5 1/2" x 9 1/4" • Pencil
The John Carter Brown Library, Brown University, Providence, Rhode Island [19]

22 Henry Cheever Pratt (Fig. 45 • Pg 100)
Moqui Village, Arizona, undated
10" x 14" • Watercolor
The John Carter Brown Library, Brown University, Providence, Rhode Island [20]

23 Henry Cheever Pratt (Plate 15 • Pg 124)
San Lazaro, Arizona, September 29, 1851
9 3/4" x 14" • Watercolor
The John Carter Brown Library, Brown University, Providence, Rhode Island [21]

24 Henry Cheever Pratt (Fig. 33 • Pg 58)
Organ Mountains from the Rio Grande, September 1852
5 1/4" x 8 1/4" • Pencil
The John Carter Brown Library, Brown University, Providence, Rhode Island [24]

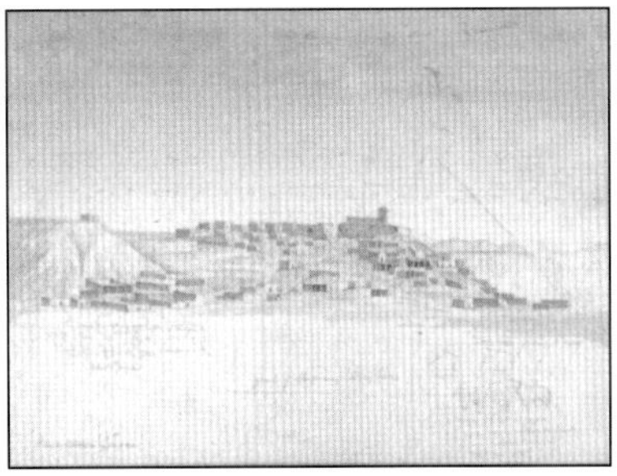

Ex. #29

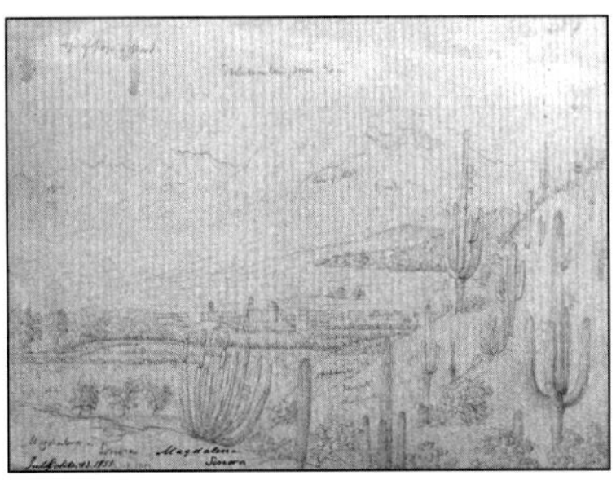

Ex. #31

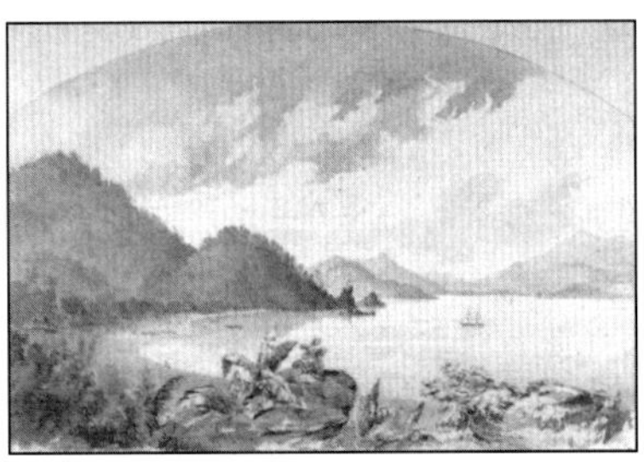

Ex. #32

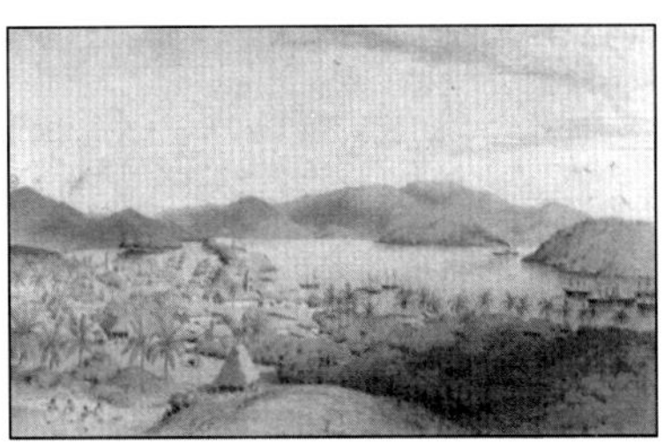

Ex. #33

25 Henry Cheever Pratt (Fig. 31 • Pg 53)
San Xavier del Bac, July 19, 1852
10" x 15" • Watercolor
The John Carter Brown Library, Brown University, Providence, Rhode Island [30]

26 John Russell Bartlett (Fig. 7 • Pg 36)
Camp in a Snow Storm on Delaware Creek, Texas, November 6, 1850
10 1/8" x 15 1/8" • Pencil and sepia wash
The John Carter Brown Library, Brown University, Providence, Rhode Island [34]

27 John Russell Bartlett (Fig. 8 • Pg. 37)
Guadalupe Mountain, Texas, November 10, 1850
9 1/2" x 13" • Pencil and sepia wash
The John Carter Brown Library, Brown University, Providence, Rhode Island [35]

28 John Russell Bartlett (Fig. 36 • Pg 60)
Giant of the Mimbres near Santa Rita, May 1, 1851
9 1/2" x 13" • Sepia wash
The John Carter Brown Library, Brown University, Providence, Rhode Island [37]

29 John Russell Bartlett
Fronteras, Sonora, May 24, 1851
9 5/8" x 13" • Pencil and sepia wash
The John Carter Brown Library, Brown University, Providence, Rhode Island [39]

30 John Russell Bartlett (Fig. 9 • Pg 37)
Guadalupe Pass, Cooke's Road, Sonora, ca. 1851
11 1/2" x 15 7/8" • Sepia wash
The John Carter Brown Library, Brown University, Providence, Rhode Island [41]

31 John Russell Bartlett
Magdalena, Sonora, October 3, 1851
9 5/8" x 13" • Pencil
The John Carter Brown Library, Brown University, Providence, Rhode Island [50]

32 John Russell Bartlett
Manzanillo on the Pacific Coast, January 1852
7 5/8" x 11 1/2" • Pencil and wash
The John Carter Brown Library, Brown University, Providence, Rhode Island [52]

33 John Russell Bartlett
Acapulco, Mexico, February 1, 1852
8 1/4" x 13 3/4" • Pencil
The John Carter Brown Library, Brown University, Providence, Rhode Island [54]

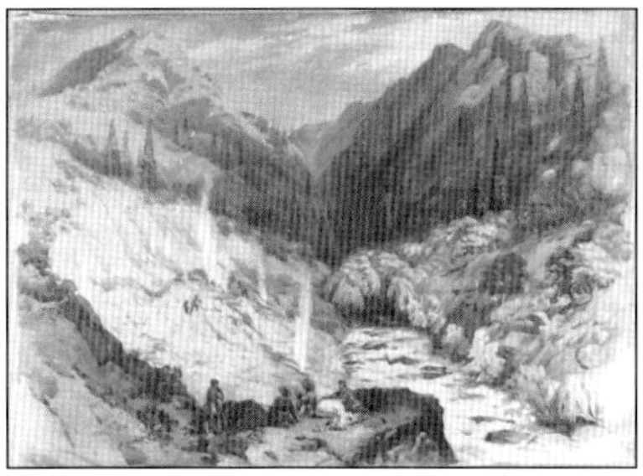

Ex. #34

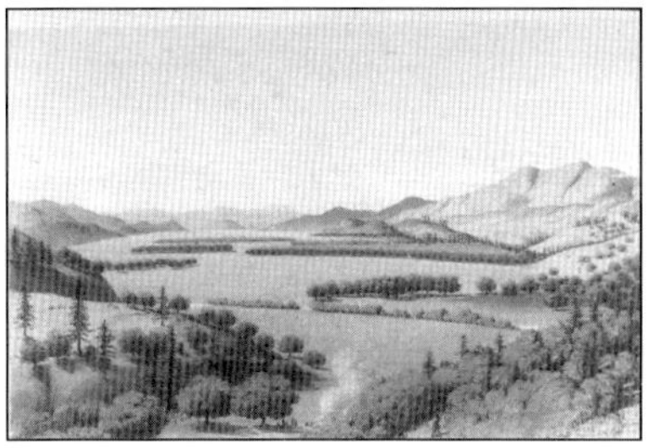

Ex. #35

34 John Russell Bartlett
Geysers-California, March 23, 1852
11 3/8" x 15 7/8" • Sepia
The John Carter Brown Library, Brown University, Providence, Rhode Island [57]

35 John Russell Bartlett
Napa Valley, California, March 19-20, 1852
9 1/2" x 13" • Pencil and wash
The John Carter Brown Library, Brown University, Providence, Rhode Island [58]

36 John Russell Bartlett (Fig. 38 • Pg 62)
Quicksilver Furnaces, New Almaden, April 4-5, 1852
11 1/2" x 16" • Pencil and sepia wash
The John Carter Brown Library, Brown University, Providence, Rhode Island [60]

37 John Russell Bartlett (Fig. 39 • Pg 63)
Ascent to the Quicksilver Mines, California, April 4-5, 1852
17 1/2" x 12 3/4" • Pencil and sepia wash
The John Carter Brown Library, Brown University, Providence, Rhode Island [61]

38 John Russell Bartlett (Fig. 6 • Pg 36)
Crossing the Pecos, October 30, 1850
9" x 12" • Pencil and sepia wash
The John Carter Brown Library, Brown University, Providence, Rhode Island [64]

39 John Russell Bartlett (Fig. 17 • Pg 40)
Initial Point, San Diego, February 19, 1852
9 3/4" x 13" • Pencil
The John Carter Brown Library, Brown University, Providence, Rhode Island [65]

40 John Russell Bartlett (Fig. 12 • Pg 38)
Big Horn Mountains, Rio Gila, June 22-23, 1852
12 3/4" x 18" • Pencil and sepia wash
The John Carter Brown Library, Brown University, Providence, Rhode Island [68]

41 John Russell Bartlett (Fig. 27 • Pg 50)
Pimo [sic] *Indian Weaving*, June - July, 1852
4 1/2" x 7 1/4" • Pencil
The John Carter Brown Library, Brown University, Providence, Rhode Island [69]

42 John Russell Bartlett (Fig. 26 • Pg 49)
Pimo [sic] *Villages & Cultivated Fields, with Desert between the Gila and the Salt* [Rivers], July 3-5, 1852
8 1/2" x 14" • Pencil
The John Carter Brown Library, Brown University, Providence, Rhode Island [72]

43 John Russell Bartlett (Fig. 30 • Pg. 52)
Tucson, Arizona, July 17-18, 1852
12 3/4" x 18 1/8" • Pencil and sepia wash
The John Carter Brown Library, Brown University, Providence, Rhode Island [75]

44 John Russell Bartlett (Fig. 44 • Pg 99)
Antiquities from the Ruins of Casas Grandes, Chihuahua, 1852
9" x 5 7/8" • Pencil and ink
The John Carter Brown Library, Brown University, Providence, Rhode Island [79]

45 John Russell Bartlett (Fig. 28 • Pg 51)
Casas Grandes, Chihuahua, August 9-10, 1852
11" x 15" • Pencil and sepia wash
The John Carter Brown Library, Brown University, Providence, Rhode Island [80]

46 John Russell Bartlett (Fig. 22 • Pg 43)
Military Post, El Paso, Texas, August 1852
8 1/8" x 11 3/8" • Sepia wash
The John Carter Brown Library, Brown University, Providence, Rhode Island [81]

47 John Russell Bartlett (Fig. 34 • Pg 58)
Organ Mountains, New Mexico, September 1852
13" x 18 1/4" • Pencil and sepia wash
The John Carter Brown Library, Brown University, Providence, Rhode Island [82]

48 John Russell Bartlett (Fig. 21 • Pg 43)
El Paso, Texas, ca. 1850-1851
8 7/8" x 11 1/4" • Sepia wash
The John Carter Brown Library, Brown University, Providence, Rhode Island [83]

49 John Russell Bartlett (Fig. 15 • Pg 39)
Camp at Small Lagoon on the Gila, June 23 or 28, 1852
8 1/2" x 14" • Pencil
The John Carter Brown Library, Brown University, Providence, Rhode Island [84]

50 John Russell Bartlett (Fig. 11 • Pg 38)
View on the Gila, 2d Camp from Ft. Yuma, June 1852
13" x 18" • Pencil and sepia wash
The John Carter Brown Library, Brown University, Providence, Rhode Island [95]

51 John Russell Bartlett (Fig. 14 • Pg 39)
On the Salinas, North of the Gila, New Mexico, July 3-4, 1852
7 1/4" x 10 3/4" • Pencil
The John Carter Brown Library, Brown University, Providence, Rhode Island [98]

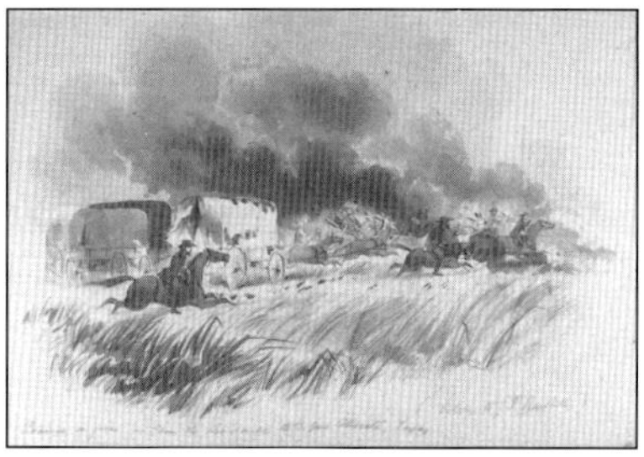

Ex. #55

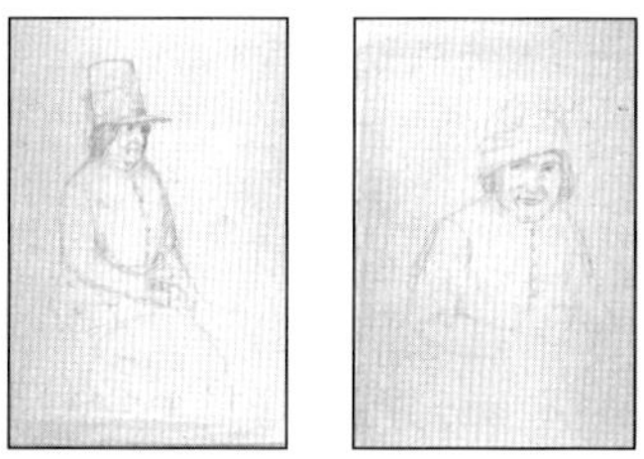

Ex. #57 a,b

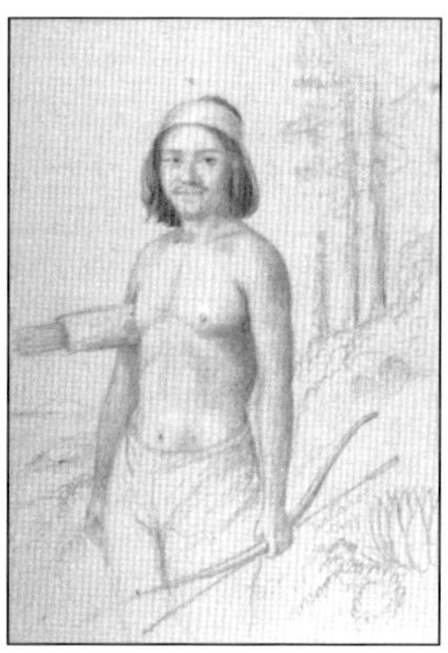

Ex. #58

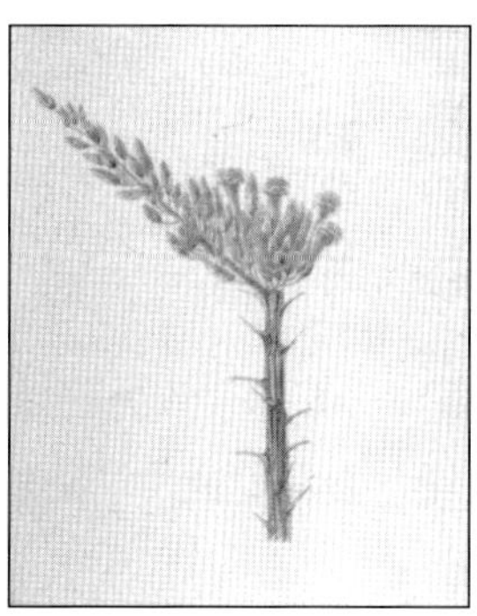

Ex. #60

52 John Russell Bartlett (Fig. 36 • Pg 60)
Granite Masses, Waco Mountains, Texas, March 29, 1851
13" x 18 1/4" • Pencil
The John Carter Brown Library, Brown University, Providence, Rhode Island [102]

53 John Russell Bartlett (Fig. 16 • Pg 40)
An Encampment, ca. 1852
9 1/8" x 14 1/4" • Pencil
The John Carter Brown Library, Brown University, Providence, Rhode Island [103]

54 John Russell Bartlett (Fig. 42 • Pg 68)
Cañon between Bacuachi and Arispe, May 30, 1851
8 1/2" x 14" • Pencil
The John Carter Brown Library, Brown University, Providence, Rhode Island [109]

55 John Russell Bartlett
Prairie on Fire- from the Rio Grande to Corpus Christi, Texas, December, 1852
10" x 12 3/4" • Pencil and sepia wash
The John Carter Brown Library, Brown University, Providence, Rhode Island [116]

56 John Russell Bartlett (Fig. 10, 26 • Pg 38, 48)
Fording a Stream, Packmules Sink in Quicksand, September 13, 1851
10" x 12 3/4" • Pencil and sepia wash
The John Carter Brown Library, Brown University, Providence, Rhode Island [117]

57 a, b John Russell Bartlett
Unidentified Indian at San Luis Rey, April 1852
13" x 18" • Pencil
The John Carter Brown Library, Brown University, Providence, Rhode Island [120a,b]

58 John Russell Bartlett
Wallahoo Chief, not dated
10 1/8" x 7 1/4" • Pencil
The John Carter Brown Library, Brown University, Providence, Rhode Island [125]

59 John Russell Bartlett (Fig. 24 • Pg 43)
El Paso, November 1850
9 3/4" x 14" • Pen and ink
The John Carter Brown Library, Brown University, Providence, Rhode Island [135]

60 John Russell Bartlett
Flowering Cactus, 1852
8 1/8" x 6 3/8" • Watercolor
The John Carter Brown Library, Brown University, Providence, Rhode Island [142]

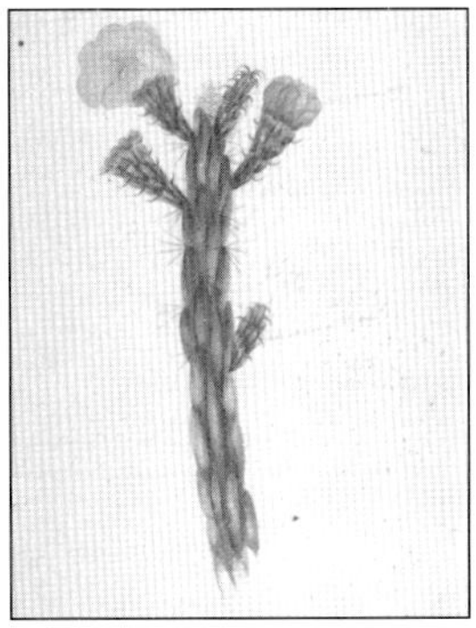

Ex. #61

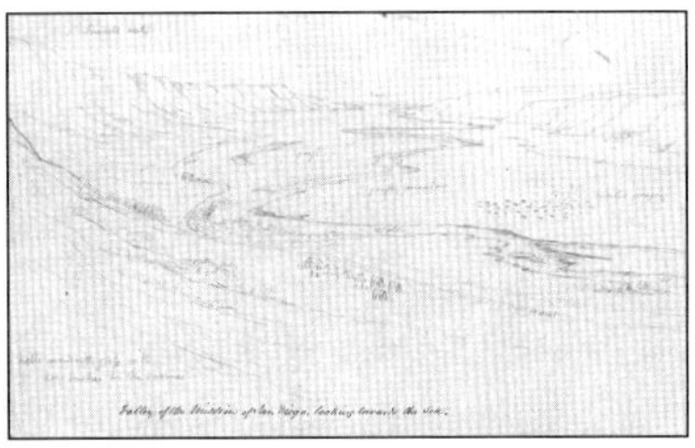

Ex. #63

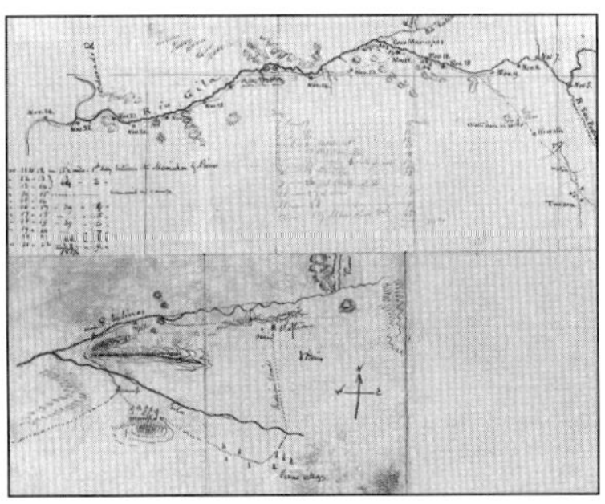

Ex. #65

Ex. #67

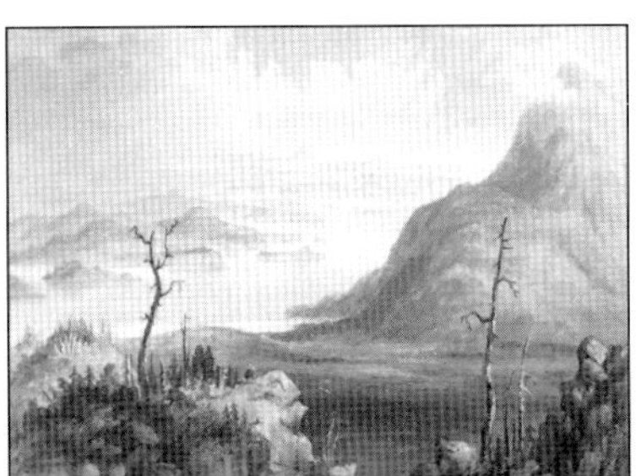

Ex. #69

61 John Russell Bartlett
Flowering Cactus (detail), ca. 1852
13" x 9 1/2" • Watercolor
The John Carter Brown Library, Brown University, Providence, Rhode Island [143]

62 John Russell Bartlett (Fig. 41 • Pg 67)
Flower of Cereus Giganteus - the Great Cactus, June-July 1852
8 1/2 x 13 1/4 • Pencil
The John Carter Brown Library, Brown University, Providence, Rhode Island [149]

63 John Russell Bartlett
Valley of the Mission of San Diego, 1852
8 1/4" x 13 1/4" • Pencil
The John Carter Brown Library, Brown University, Providence, Rhode Island [153]

64 John Russell Bartlett (Fig. 46 • Pg 105)
Sculptured Rocks near the Gila, 1852
8 1/2" x 11 1/2" • Sepia wash
The John Carter Brown Library, Brown University, Providence, Rhode Island [154]

65 John Russell Bartlett
Map of the Gila River, ca. 1852
8 1/4" x 10" • Pencil
The John Carter Brown Library, Brown University, Providence, Rhode Island [164]

66 Henry Cheever Pratt (Fig. 5 • Pg 35)
Portrait of John Russell Bartlett, ca. 1852
10" x 7 1/4" • Pencil
The John Carter Brown Library, Brown University, Providence, Rhode Island [166]

67 John Russell Bartlett
Sketchbook - Los Angeles, April 1852
9" x 11 1/2" • Pencil
The John Carter Brown Library, Brown University, Providence, Rhode Island [171f]

68 John Russell Bartlett (Fig. 35 • Pg 59)
Sketchbook - Gila, 1852
9" x 12" • Pencil
The John Carter Brown Library, Brown University, Providence, Rhode Island [173]

69 Harrison Eastman
View Across the Boundary Line from Sugar Loaf, not dated
12 1/4" x 16 1/2" • Watercolor
The John Carter Brown Library, Brown University, Providence, Rhode Island [195]

70 John Russell Bartlett (Fig. 29 • Pg 52)
Sketchbook - Tucson, Sonora from the SW, July 18, 1852
9" x 12" • Pencil
The John Carter Brown Library, Brown University, Providence, Rhode Island [172w]

71 Augustus de Vaudricourt (Fig. 4 • Pg 33)
Portrait of John Russell Bartlett, ca. 1850
7 1/2 x 9 • Pencil
The John Carter Brown Library, Brown University, Providence, Rhode Island

72 Henry Cheever Pratt (Plate 14 • Pg 123)
The Deserted Mission of Tumacacori, ca. 1854-1855
30" x 46" • Oil on canvas
Loan courtesy of Don Heald, New York, New York

73 Henry Cheever Pratt (Plate 22 • Pg 131)
Allegorical Design Representing Evening with a View from Nature Painted on the Gila River, Arizona, ca. 1853
36" x 46" • Oil on canvas
Loan courtesy Collection of Philip Y. De Normandie

74 Henry Cheever Pratt (Plate 5 • Pg 114)
Portrait of James Wiley Magoffin, October 1852
36" x 28 3/4" • Oil on canvas
Loan courtesy of Magoffin Home State Historic Site, El Paso, Texas

75 Attributed to Henry Cheever Pratt (Plate 13 • Pg 122)
San Xavier del Bac Mission Church, after 1860
25" x 29" • Oil on canvas
Loan courtesy of Arthur J. Phelan Jr.

76 Henry Cheever Pratt (Plate 19 • Pg 128)
Bartlett Party Traversing a Canyon Bewteen Bacuachi and Arispe, 1855
18" x 24" • Oil on canvas
Loan courtesy of Phoenix Art Museum from the Carl S. Dentzel family collection, Phoenix, Arizona

77 Henry Cheever Pratt (Plate 18 • Pg 127)
Basin of the Rio Gila, 1855
26 1/4" x 36 1/16" • Oil on canvas
Loan courtesy of Phoenix Art Museum from the Carl S. Dentzel family collection, Phoenix, Arizona

78 Seth Eastman (Plate 20 • Pg 129)
Cañon Leading to Magdalena, Sonora, Mexico, ca. 1853
9 1/2" x 12 1/4" • Watercolor and pencil
Loan courtesy of Museum of Art, Rhode Island School of Design, Providence [47.112.1]

79 Seth Eastman (Plate 34 • Pg 143)
Camping in a Storm on the Mexican Plateau, ca. 1853
9 1/8" x 12 1/2" • Watercolor and pencil
Loan courtesy of Museum of Art, Rhode Island School of Design, Providence [47.112.2]

80 Seth Eastman (Plate 31 • Pg 140)
Great Cañon, River Gila, ca. 1853
12 1/2" x 9 5/8" • Watercolor and pencil
Loan courtesy of Museum of Art, Rhode Island School of Design, Providence [47.112.9]

81 Seth Eastman (Plate 21 • Pg 130)
Bufa del Cobra [sic], *Rocky Mountains*, ca. 1853
9 1/4" x 12 1/4" • Watercolor and pencil
Loan courtesy of Museum of Art, Rhode Island School of Design, Providence [47.112.3]

82 Seth Eastman (Plate 8 • Pg 117)
Village of the Pimo [sic] *Indians, Gila River*, ca. 1853
9 1/8" x 12 1/8" • Watercolor and pencil
Loan courtesy of Museum of Art, Rhode Island School of Design, Providence [47.112.11]

83 Seth Eastman (Plate 29 • Pg 138)
View from the Camp near the Boundary Line 4th day from Fort Webster, ca. 1853
9 1/2" x 12 1/2" • Watercolor and Pencil
Loan courtesy of Museum of Art, Rhode Island School of Design, Providence [47.112.4]

84 Seth Eastman (Plate 28 • Pg 137)
Santa Rita del Cobra [sic], *Rocky Mountains, New Mexico*, ca. 1853
9 1/2" x 12 1/2" • Watercolor and pencil
Loan courtesy of Museum of Art, Rhode Island School of Design, Providence [47.112.6]

85 Seth Eastman (Plate 30 • Pg 139)
Pass at the Pitoncillo, [sic] *Chiricahui Mountains in the Distance*, ca. 1853
9 1/4" x 12 3/8" • Watercolor and pencil
Loan courtesy of Museum of Art, Rhode Island School of Design, Providence [47.112.5]

86 Seth Eastman (Plate 27 • Pg 136)
Cañon Leading to the Copper Mines at Santa Rita del Cobra [sic], ca. 1853
9 3/8" x 12 1/2" • Watercolor and pencil
Loan courtesy of Museum of Art, Rhode Island School of Design, Providence [47.112.]

87 Seth Eastman (Plate 33 • Pg 142)
Vallecito, California, ca. 1853
9 3/8" x 12 3/8" • Watercolor and pencil
Loan courtesy of Museum of Art, Rhode Island School of Design, Providence [47.112.7]

88 Seth Eastman (Plate 35 • Pg 144)
Camp near Lake de Los Patos, Chihuahua, ca. 1853
9" x 12 3/8" • Watercolor and pencil
Loan courtesy of Museum of Art, Rhode Island School of Design, Providence [47.112.8]

89 Seth Eastman (Plate 26 • Pg 135)
Fording Devil's River, Texas, ca. 1853
8 3/4" x 12" • Watercolor and pencil
Loan courtesy of Museum of Art, Rhode Island School of Design, Providence [47.112.10]

90 Seth Eastman (Plate 32 • Pg 141)
Santa Cruz Valley, ca. 1853
9 3/8" x 12 1/4" • Watercolor and pencil
Loan courtesy of Museum of Art, Rhode Island School of Design, Providence [47.112.13]

91 Henry Cheever Pratt (Plate 10 • Pg 119)
View from Maricopa Mountain near the Rio Gila, ca. 1855
33" x 48" • Oil on canvas
Loan courtesy of The Rosewood Corporation, Dallas, Texas

92 James Smillie (not shown in catalogue)
The Voyage of Life: Youth, ca. 1848
18 7/8" x 25" • Lithograph
Loan courtesy of The University of New Mexico Art Museum, Albuquerque, New Mexico

93 Henry Cheever Pratt (Plate 7 • Pg 116)
Coons Rancho, Paseo del Norte, Texas, 1851-1852
32" x 50" • Oil on canvas
Loan courtesy of Texas Memorial Museum, University of Texas, Austin Texas

THE ALBUQUERQUE MUSEUM

Printing and Prepress, Guynes Lithographers
Cover Prepress, Universal Color
Text: Goudy Oldstyle, and Carpenter designed in PageMaker and Adobe Illustrator